# Waste to Strength: Building Green Concrete

**Namita**

# Table of Contents

# Chapter 1   Introduction

The environmental concerns about the production of cement in terms of the carbon dioxide ($CO_2$) emission lead to the search for more by-product materials as alternatives to cement. Geopolymer Concrete (GPC) is one of those alternative materials where waste materials such as fly-ash and slag are totally replaced with cement. Use of fly-ash as one of the raw materials minimize the production of waste materials and protects the environments. However, bottom-ash as another waste material in coal-fired power plants has the potential to be used as a replacement or addition to fly-ash in producing GPC. Limited studies were performed on the mechanical properties of GPC made by both fly-ash and bottom-ash. Hence, the main purpose of this research is to provide a thorough understanding of workability and durability of GPC made by fly-ash and bottom-ash including freeze-thaw resistance, leach-ability, and dynamic elastic modulus in both laboratory and real environmental conditions.

## 1.1   Research motivation

Portland Cement Concrete's (PCC) versatility, durability, and economy have made it the world's most extensively used construction materials. However, PCC is considered as an unsustainable material because the manufacture of cement as a main raw material of PCC releases a huge amount of $CO_2$ into the atmosphere [1]. Moreover, the coal-based thermal power plant produces by-products approximately 112 million tons annually [2]. This amount of by-products needs a large area of land for disposal. Due to the abovementioned issues, over the past decades, attempts have been made to replace cement with by-products and use them as the main constituent for production of concrete to increase the recycled content in building materials and reduce the $CO_2$ emission.

GPC is known as a sustainable alternative material to PCC that is produced by mixing the by-products and alkaline activators. The concept of "Geopolymer" was initially introduced by a French scientist, Joseph Davidovits. Then, slag-based geopolymer cement was made in the 1980s [3]. Afterwards, in 1997, based on the obtained results for slag-based geopolymer cement, Silverstrim et al [4], Van Jaarsveld et al. [5] created a fly-ash based geopolymer cement. Since then, most of the previous studies on GPC are about the mechanical properties of fly-ash based GPC [6-8]. However, bottom-ash is quite similar to fly-ash in terms of physical and chemical components. As a result, several trial experiments have been made to make GPC with optimum durability substituting 50% bottom-ash with 50% fly-ash.

Generally, a concrete structure with the open surface is under the risk of frost at cold climates because the water inside the pores get frozen and causes volume expansion of the structure. Once there is no free space for unfrozen water in air voids/pores of the concrete, the unfrozen water flows through the capillary and gel pores. These phenomena create capillary and osmotic pressure. So, these pressures cause the initiation of cracks and spalls under the effect of freeze-thaw cycles. Generally, concrete has relatively large capillary pores and air voids and a relatively large amount of free water inside. For saturated concrete, only its capillary and gel pores may be filled with water. The size of the capillary pores inside saturated concrete is generally within 0.5 $\mu m$, and the water inside the capillary pores only freezes below -12 °C. In contrast, the water inside the gel pores inside saturated concrete freezes only below -78 °C and is unfrozen under natural conditions. Therefore, this water will not cause freeze-thaw damage to the concrete. In the actual freeze-thaw process, the freeze-thaw damage in concrete is caused by the hydrostatic pressure generated as a result of freezing the water inside the capillary pores, i.e., hydrostatic pressure is the driving force for freeze-thaw damage in concrete and based on the ASTM C 666 [9], the nominal freeze-thaw cycle shall consist of alternately lowering the temperature of the specimens from 4 to -18 °C and raising it from -18 to 4 °C in not less than 2 nor more than 5 hours. So, this standard is considered the temperature higher than actual freeze- thaw process in capillary pores. Hence, it is essential to increase the freeze-thaw durability of the GPC structure [10,11]. To overcome internal degradation due to the freeze-thaw cycles, there must be enough free space to accommodate the volume expansion and reduce the tension pressure. As a solution to this issue, Air Entraining Admixture (AEA) is used to create microscopic air bubbles in the concrete, which improve the durability of concrete when subjected to freeze-thaw conditions [12-14]. In this study, due to the research gap on freeze-thaw durability of bottom-ash based

GPC, GPC and PCC paver blocks were exposed to real environmental conditions to make the comparison between service-life of these two types of concrete.

The current movement toward using waste materials in concrete production increases the risk of pollution to ground-water and atmosphere because most of the waste materials such as fly-ash and bottom-ash contain different toxic heavy metals [15]. Hence, it is critical to perform further work to determine the leach-ability of K-based GPC to prevent environmental pollution. In this study, the leach-ability of GPC was measured at the laboratory and in real environmental condition. The Toxicity Characteristic Leaching Procedure (TCLP) as a laboratory and Non-Destructive Technique (NDT) was used to analyze all the toxic metals in GPC such as Cd, Cu, Zn and Pb. The leaching test strips as NDT/field method was used to measure the effect of real environmental conditions on leach-ability of GPC and PCC.

## 1.2    Dissertation outline and objectives

In **Chapter 1** of this dissertation the context and framework are provided. Furthermore, this chapter proposes the interaction between the following Chapters.

**Chapter 2** summarizes the literature review carried out on different properties of GPC. The objective of this section is to identify gaps in the current research knowledge on GPC and to understand the effect of laboratory and real environmental conditions on durability and mechanical properties of GPC. More overview of GPC along with its processing parameters are discussed in other chapters.

In **Chapter 3** of this dissertation the methodologies are provided.

**Chapter 4** is a published journal manuscript (peer-reviewed) that presents the final mix proportion of GPC, using two curing methods after manufacturing of different trial mix proportions/samples. As a meso-scale study, the objective of this section is to find a superior curing method to achieve higher compressive strength. Hence, GPC specimens were cured (dry and steam) at different temperatures including ambient, 30 °C, 45 °C, 60 °C and 80 °C for 24 hours. The compression test was performed on the GPC specimens to find the highest strength. A wide range of mix proportions was made for Na-based GPC using fly-ash, but limited work has been performed on K-based GPC using 50% bottom-ash and 50% fly-ash. So, this work shows the feasibility of designing sustainable concrete using both fly-ash and bottom-ash, employing alkali-activation technology (K-based). As another objective of this study, the resonant frequency of GPC and PCC was measured to investigate the dynamic elastic modulus. Finally, attempts have been made to use different proposed predictive methods to find their applicability for GPC.

**Chapter 5** is a novel peer-reviewed published journal manuscript that presents the microstructure of GPC air voids in a frozen condition using a 4-D Low Temperature Scanning Electron Microscopic (4-D LTSEM) device. Freeze-thaw cycles in a frozen condition lead to the initiation of cracks and spalls due to progressive expansion of the paste. Consequently, the structure loses its durability. Since there is research gap on the investigation of cryo formation of GPC due to the unavailability of 4-D LTSEM device, the main objective of this Chapter was microstructurally investigating the morphology of cryo formation inside the air voids of GPC paste. Moreover, most of the models and standards for GPC are proposed based on available information for PCC. Therefore, another objective of this study is to have a comprehensive understanding of the freezing point of GPC.

**Chapter 6** is a published journal manuscript (peer-reviewed) that presents the freeze-thaw resistance of GPC under controlled conditions. As a meso-scale study, Fly-ash based GPC and bottom-ash GPC specimens were tested under accelerated freeze-thaw cycles. The dynamic elastic modulus of GPC specimens was measured every 30 cycles. The main objective of this work was to make a service-life prediction model based on the results of the freeze-thaw test and dynamic elastic modulus. Another objective of this Chapter was to trace all heavy metals that may leach from the GPC specimens and impact on the environment. Since there is no work reported on leach-ability of bottom-ash based GPC, it was essential to compare the obtained results with an available standard.

**Chapter 7** presents a published journal manuscript (peer-reviewed) on corrosion resistance of GPC, using two NDT techniques (half-cell potential and linear polarization resistance). As mentioned above, corrosion is another type of damage that leads to deterioration of structure. So, due to the research gap in literatures for this type of concrete, attempts have been made to experimentally compare the corrosion resistance of GPC and PCC using accelerating methods.

**Chapter 8** is a published journal manuscript (peer-reviewed) that shows the durability and leach-ability of GPC and PCC paver blocks in real environmental conditions. As a macro-scale study, two small pieces of the site were paved. One site was considered to investigate the effect of weathering on leaching of paver blocks and another site was considered to investigate the effect real environmental conditions on the durability of paver blocks. Three NDTs including Schmidt hammer, Ultrasonic Pulse Velocity (UPV), and leaching strips were used to make a correlation between dynamic elastic modulus and compressive strength over 240 days of exposure. The main objective of this Chapter was to investigate durability and heavy metal leach-ability of GPC paver blocks in real environmental conditions.

In **Chapter 9**, the flexural strength and traffic analysis and steam curing cost of GPC are only measured and briefly explained.

**Chapter 10** summarizes the main results and contributions of the entire work done in this project and suggests the possible future work.

### 1.3  Research contributions

As the demand for concrete as a construction material increases, hence the demand for cement does, too. However, cement production releases a huge amount of $CO_2$ into the environment. So, the development of green materials having low energy consumption, greenhouse gas emission and zero waste must be overemphasized. Moreover, increasing the lifespan/strength of GPC in cold regions has both environmental and economic benefits. So, during this study, attempts have been made to produce sustainable concrete (GPC) with optimum workability and durability including freeze-thaw resistance.

Lack of information on workability and durability of GPC in real environmental conditions has become apparent. So, in comparison to previous studies, in a larger scale, this work aimed to systematically study the durability (such as freeze-thaw resistance) and chemical metals leach-ability of GPC paver blocks when they were exposed to a harsh condition. In addition to addressing this research, authors could not find any freeze-thaw damage theory using dynamic elastic modulus for bottom-ash based GPC. Hence, a predictive model for paver blocks is also proposed based on their dynamic elastic modulus. This predictive model can be used as an academic tool in designing sustainable GPC when exposed to a cold climate. Furthermore, since waste materials are used in the mixture of this type of concrete, leach-ability of GPC is investigated in both real environmental and laboratory conditions.

Moreover, it is vital to capture the effect of freeze-thaw cycles on the microstructure of concrete. However, there is no information reported on cryo formation inside the air voids of GPC because most of the Scanning Electron Microscopic (SEM) equipment cannot maintain the low temperature required to capture an image from a frozen sample. That is why, as novel research, 4-D LTSEM device as a rare equipment was used to study the microstructure of air voids of frozen GPC samples. This study contributes to investigating applicability of ASTM C666 [9] for GPC made by fly-ash and bottom-ash. Moreover, this study leads to gain better ideas about the morphology of GPC in a frozen condition and consequently, creating a paradigm for future studies of the evolution of durability (freeze-thaw resistance) of GPC.

The corrosion of steel rebar in GPC is another main reason for the concrete delamination. Using NDT for measuring corrosion rate of GPC is an essential requirement for the prediction of damage regions. Thus, the corrosion rates of GPC and PCC were compared using half-cell potential and linear polarization resistance techniques. This comparison is also beneficial for inspectors due to lack of knowledge on the accuracy of corrosion NDTs. This part of the dissertation was already presented as a graduate project at the University of Victoria by Morla Priyanka and Dr. Rishi Gupta [16]. The contribution of the current candidate is explained later.

In a nutshell, this research brings superior advantages over conventional concretes. In fact, manufacturing GPC uses the lowest $CO_2$ than the manufacturing of Portland cement does. It is long-lasting than normal concrete and needs little resources, thus saves huge amounts of budgets that would otherwise be spent on the production of cement and environmental impacts.

# Chapter 2   Literature Review

This chapter summarizes the literature review carried out on various properties of GPC to identify gaps in the previous studies. GPC has become a very popular subject of research in recent years because GPC has the potential of entirely replacing cement-based concrete (due to $CO_2$ emission) with waste materials [17]. GPC made by only fly-ash or slag has indicated excellent mechanical properties and durability [18,19]. However, the mechanical properties and durability of GPC made by both fly-ash and bottom-ash in a cold condition have been studied by a few researcher. That is why, in this chapter, attempts have been made to focus on the durability of GPC made by fly-ash and bottom-ash in frozen conditions since the main objectives of this study are freeze-thaw resistance and leach-ability of this type of concrete in both real environmental conditions and lab condition. It should be noted that there are more details on the literature review in later chapters.

## 2.1   Constituents of GPC and their effects

GPC is produced by the reaction of aluminosilicate powder with an alkaline solution. Once aluminosilicate solids are dissolved, and once $[SiO_4]-$ and $[AlO_4]-$ tetrahedral units are released in solution, polymerization takes place under alkaline conditions [20,21].

$$(Si_2O_5Al_2O_2)_n + H_2O + OH- \rightarrow Si(OH)_4 + Al(OH)^{4-} \qquad (1)$$

In the last decade, fly-ash based GPC has been considered as a cement-based concrete alternative in the construction industry. Fly-ash based GPC has shown excellent properties such as compressive strength, low creep and low shrinkage [22-24]. Bottom-ash is another part of the non-combustible residue of combustion in a power plant, boiler, furnace or incinerator. Bottom-ash is the coarse, granular, incombustible by-product of coal combustion that is collected from the bottom of furnaces. Furthermore, the chemical compositions of bottom-ash are similar to that of fly-ash. There are numerous studies replacing cement with fly-ash to create concrete. However, due to limited utilization of bottom-ash, dealing with bottom-ash is now posing to be one of the most important challenges in recent decades. So, in this study, a mix proportion for potassium-based GPC using combination of fly-ash and bottom-ash was made to seek after quality advancement of bottom-ash based GPC that can be utilized as building material in future. Tianyu et al. [25] investigated compressive strength, elastic modulus, flexural strength, workability, drying shrinkage and absorption capacity of fly-ash based, bottom-ash based, and blended fly-ash and bottom-ash based GPC. The achieved results showed that the durability of GPC specimens increases with a decrease in the liquid-to-binder ratio or an increase in the mass ratio of the fly-ash to bottom-ash. However, in their study, Na-based solution was used in the production of GPC. Hence, in the current study, attempts have been made to create a mix proportion (target strength of 35 MPa) after trial experiments such as compressive strength on K-based GPC synthesized with the combination of 50% fly-ash and 50% bottom-ash.

As mentioned before, the alkaline solution also affects the durability of GPC. Previously, Na-based alkaline solution was used to produce GPC [26-28]. It was reported that a Na-based solution can liberate more silicate and aluminate monomers than K-based solution [29]. Because sodium cations, which are smaller than that of the potassium cations, enables cations to move easily in the paste matrix. However, the K-based solution can also be used to begin the geopolymerization process. The K-based solution was used in the current study because at elevated temperature (higher than 30 °C), GPC made by K-based is more steady than Na-based GPC in terms of mechanical properties including compressive strength [30]. Moreover, in the current study, K-based alkaline solution is used to develop a durable construction material, since limited work is performed on K-based GPC.

## 2.2 Curing conditions

Numerous attempts have been made to investigate the effect of the curing period and temperature on the durability of GPC [31-33]. Ebrahim et al. [34] produced three mix proportions (oven-cured) at different temperatures (45 °C, 65 °C, 85 °C) and time (5–20 hours) after 1 and 7 days of procuring. The results showed that optimum curing temperature for GPC samples is essential to attain optimal strength. However, other types of curing methods such as steam curing were not used in their study. Therefore, in the current research, three curing regimes including ambient, dry and steam were used to investigate the effect of curing method on various properties of K-based GPC. Palomo et al. [22] investigated the effect of curing on fly-ash based GPC at 65 °C and 85 °C. The results showed that the compressive strength of GPC samples cured at 85 °C for 24 hours, that was higher than those samples cured at 65 °C. However, in their study, only fly-ash was used as an alumina-silicate material. Hence, the effect of curing regime on mechanical properties of GPC made by both fly-ash and bottom-ash should be investigated due to the limited work on this type of mix proportion. Heah et al. [33] proposed that the strength of metakaolin-based GPC samples cured at ambient temperature was lower than heat-cured samples (40 °C, 60 °C, 80 °C, 100 ° C) after 1–3 days. However, curing at a higher temperature for a longer period caused the failure of samples at a later age due to the thermolysis of –Si–O–Al–O– bond. However, the development of K-based GPC made by fly-ash and bottom-ash in terms of curing regime has not yet been studied. Therefore, in this study, K-based GPC samples made by fly-ash and bottom-ash were cured at different temperatures including ambient, 30 °C, 45 °C, 60 °C and 80 °C using three types of curing method (ambient, steam and dry) to obtain higher compressive strength.

In another study, Kovalchuk et al. [35] investigated the effect of curing conditions on fly-ash based GPC. Three types of curing methods were used to produce GPC including covered moulds (8 hours at 95 °C), dry curing (first, moulds were kept in an oven for 2 hours at 95 °C and then, the specimens were demoulded and kept in an oven at room temperate to 150 °C in 3 hours and 6 hours), and steam curing (first, the moulds were kept in an oven for 2 hours at 95 °C and then, the demoulded samples were exposed to the following thermal ramp: at room temperature to 95 °C in 3 hours, 6 hours at 95 °C, and cooling to ambient temperature in 3 hours). According to the results, covered moulds reached the highest compressive strength of 102 MPa. Therefore, in the current study, the aim was to find the curing regime that optimizes the polymerization reactions taking place during the K-based alkali activation of fly-ash and bottom-ash, using different curing methods and temperatures.

## 2.3 Relative dynamic modulus of elasticity

The elastic modulus is an important mechanical property of GPC when the structure is under compression stress. Thus, it is important to study the elastic modulus of GPC and compare it with conventional concrete. Yang et al. [36] investigated the elastic modulus of PCC and fly-ash based GPC. They observed that with similar compressive strength (about 30 MPa), fly-ash based GPC has 2000 Hz lower resonant frequency than PCC, which led to the lower elastic modulus. However, there are limited studies on the modulus of elasticity of GPC made using bottom-ash. The authors of the current study also could not find any study that predicts and compares the elastic modulus of this type of mixture (bottom-ash based GPC) with PCC. In another study, Ana et al. [37] studied the engineering properties of alkali-activated fly-ash concrete and PCC including static modulus of elasticity. Two mix proportions were made to produce Na-based alkali activated concrete. Samples were oven-cured 20 hours at 85 °C for alkali activated concrete, and 20 hours at 22 °C and 20 hours at 40 °C and 98% relative humidity for PCC. According to the results, alkali activated concrete had a lower modulus of elasticity than PCC. However, there are several studies on the static elastic modulus of GPC. That is why, in the present study, the dynamic elastic modulus of GPC made by fly-ash and bottom-ash was measured using the new technique, because limited work is reported on the dynamic elastic modulus of this type of GPC. Khalil et al. [38] investigated the dynamic elastic modulus of Na-based GPC. Several trial mixes were produced to achieve a reference GPC mix. The resonance frequency method was used to determine the dynamic elastic modulus of GPC specimens according to ASTM C215 [39]. The results showed that the elastic modulus of GPC is between 23.86 GPa and 29.86 GPa. Moreover, the modulus of elasticity increased linearly to the square root of compressive strength. In the present study, attempts have been made to experimentally find a better predictive model for GPC using the proposed empirical formula by other researchers since the current authors have not found any other studies reported in the literature on correlating elastic modulus and compressive strength of K-based GPC made using bottom-ash and fly-ash.

## 2.4 Freeze-thaw resistance and mass loss

One of the main objectives of this study is to evaluate freeze-thaw resistance of GPC, because exposure to extreme temperatures such as frost conditions is one of the key causes of structural deterioration. Zhao et al. [40] investigated the freeze-thaw resistance of fly-ash based GPC. Three types of GPC with different slag contents (GPC-10, GPC-30 and GPC-50) and PCC were produced. GPC-10 specimens were cured at 80 °C for 24 hours. The standard conditions (20±2 °C, relative humidity ≥95%) were used to cure GPC-30 and GPC-50 specimens. The dynamic elastic modulus and mass loss of GPC specimens were tested every 25 cycles of freeze-thaw. According to the results, GPC-10 showed a high rate of mass loss than other specimens. The test of GPC-10 and GPC-30 specimens were terminated after 25 freeze-thaw cycles due to the internal and external damage (scaling). The RDME of GPC-50 specimens decreased linearly over 100 cycles of freeze-thaw. The RDME of GPC-50 dropped to 40% after 125 freeze-thaw cycles. The RDME of the PCC decreased to 99.0% after 125 freeze-thaw cycles showing higher freeze-thaw resistance. However, limited work has been performed on freeze-thaw resistance of K-based GPC made by 50% fly-ash and 50% bottom-ash. That is why, this project evaluated the freeze-thaw durability of fly-ash based GPC and bottom-ash based GPC in accordance with ASTM C666 [9]. Degirmenci et al. [41] investigated the Na-based GPC specimens cured at room temperature of 23±2 °C until the testing day. In this test, the specimens were frozen at -20°C for 4 hours during the freezing period and in the water at room temperature for 4 hours during the thawing period. Three types of mixes were made (fly-ash based, slag-based and zeolite-clinoptilolite-based). The mass loss and compressive strength of specimens were tested after 25 freeze-thaw cycles. The result showed that zeolite-clinoptilolite-based GPC showed a higher rate of mass loss and strength loss after 25 freeze-thaw cycles compared to other samples. However, in the current study, accelerating freeze-thaw machine was used to expose two mix proportions to 300 freeze-thaw cycles including fly-ash based GPC and bottom-ash based GPC. Moreover, attempts have been made to find a predictive model when GPC is exposed to freeze-thaw conditions, since limited work has proposed a predictive model for this type of GPC.

## 2.5    Corrosion of steel reinforcement

In this study, NDTs were used to detect the state of embedded steel in GPC made by only fly-ash or slag. This is important because the corrosion of steel causes the deterioration of the structure. Numerous studies were reported to evaluate the corrosion potential of steel within the GPC [42-44]. Olivia et al. [45] investigated the corrosion potential of fly-ash based GPC and PCC. In their study, time to failure was explained as the onset time of current increase. Half-cell potential results indicated that after 91 days, GPC specimens had low-level of corrosion activity, and GPC times to failure under accelerated corrosion test were 3.86–5.70 times longer than those of the PCC specimens. Reddy et al. [44] compared the durability of GPC with that of PCC exposed to the marine environment for 21 days. According to the results, the initial corrosion current measured for GPC specimens (71–91 mA) was lower than that of PCC specimens (772 mA). The PCC specimens initially noted a decrease in the current but later started increasing. While, in GPC specimens the current never indicated a significant increase. Regarding the corrosion of reinforcement in K-based GPC made by fly-ash and bottom-ash, the literature is relatively limited. Therefore, two types of NDT were used to evaluate corrosion rate in laboratory conditions.

## 2.6 Leach-ability of GPC

It is well-known that the use of alternative waste materials in concrete production may increase contamination, especially for the aquatic environment when concrete is in contact with water. Such concretes raise doubts due to the impact of waste materials on the environment. Many environmental studies have been performed on leach-ability of conventional concrete/cement-based concrete [46- 48]. However, limited work is reported on leach-ability of GPC. Arioz et al. [49] studied the leaching of fly-ash based GPC. Fly-ash was activated by 12 molar Na-based solutions. GPC specimens were cured at 40 °C, 80 °C and 120 °C for 6, 15 and 24 hours, respectively. The TCLP method was applied for tracing heavy metals of fly-ash based GPC. The TCLP results indicated that heavy metals like As and Hg immobilized in the structure effectively. However, in the current study, the K-based solution was used to activate geopolymerization process. Hence, it was crucial to investigate the heavy metals leach-ability of K-based GPC, since the current authors could not find any work on this type of GPC. Yunsheng et al. [50] studied heavy metal immobilization behaviours of slag-based geopolymer. Four different slag contents (10%, 30%, 50%, 70%) and three types of curing methods (standard curing, steam curing and autoclave curing) were used to obtain the higher durability. According to the results, leaching tests indicated that slag-based geopolymer can effectively immobilize heavy metals like Cu and Pb. The rate of immobilization exceeds 98.5% once the amount of heavy metals is in the range of 0.1–0.3% by mass of binders. In their study, only Cu and Pb were considered for testing GPC made by fly-ash and slag. Therefore, attempts have been made to investigate the leach-ability of most of the heavy metals of K-based GPC in accordance with USEPA 1311 using TCLP method, since some controversy persists.

## 2.7  Mechanical properties

Kolli et al. [51] compared the mechanical properties of GPC and PCC. They produced seven Na-based GPC made by fly-ash mix proportions and three PCC mix propositions. The results showed that GPC has reached its target strength earlier under heat-cured condition compared to room temperature. Moreover, heat-cured GPC samples showed the same tensile strength as PCC samples. Furthermore, the mechanical strength of PCC samples was similar to that of GPC samples. However, using a new raw material such as bottom-ash plays a vital role in the geopolymerization process and affects the durability and microstructure of the final products. Hence, attempts have been made to characterize the properties of K-based GPC made by fly-ash and bottom-ash for potential practical applications.

Jian et al. [52] investigated the compressive strength of blended GPC made by Red Mud (RM) and Rice Husk Ash (RHA) at the age of 60 days. RHA was used in two conditions including ''as-received'' condition and 100% particles passing through a #100 mesh sieve. According to the results, the RHA-based GPC specimens showed a compressive strength of 16.08 MPa, an increase of 37.43% compared to equivalent unground RM-based GPC specimens. They attributed the improvement in compressive strength to the higher degree of geopolymerization process due to the existence of fine particle size and the high specific surface area of ground RHA. However, other mechanical properties and durability of GPC such as freeze-thaw resistance, elastic modulus and corrosion etc. were not measured in their study.

In another study, Sata et al. [53] studied the effect of three different particle sizes (15.7 μm, 24.5 μm and 32.2 μm) of bottom-ash on compressive strength of GPC. The GPC specimens were tested at age of 7, 28, 90, 180 and 360 days. The results showed that the finer particles led to higher compressive strength at all curing days. Nazari et al. [54] evaluated the effect of particle distribution (75 μm and 3 μm for fly-ash, and 90 μm and 7 μm for RHA) on compressive strength of GPC made by fly-ash and RHA. The GPC specimens were tested at age of 7 and 28 days. The results showed that GPC specimens with higher amount of fine fly-ash and RHA had higher compressive strength. They attributed the capability of finer particles to fill pores and create a denser paste. Since there is the limited available industrial application for GPC containing bottom-ash and due to the bottom-ash's similarity in silica and alumina content with fly-ash, the current authors have started to incorporate bottom-ash in GPC production.

Nath et al. [55] investigated the flexural strength and the elastic modulus of fly-ash based GPC exposed to ambient temperature. GPC and PCC beams (100×100×400 mm) were cast and cured at ambient temperature and in water (23±2 °C), respectively. The flexural strength test was performed at 28 and 90 days to compare the experimental results with predictive equations of different standards including AS 3600 [56] and ACI 318 [57]. According to the results, GPC showed greater flexural strength than PCC with comparable compressive strength. The flexural strength of fly-based GPC cured at ambient temperature was suggested to be modelled by AS 3600 [56] predictive equation. A comparative study of modulus of elasticity and compressive strength of Na-based GPC and PCC was also conducted by authors. Both compressive strength and modulus of elasticity of cylindrical samples were tested at 28 and 90 days. GPC showed 25%-30% less modulus of elasticity than PCC with similar compressive strength. However, the effect of bottom-ash, K-based alkali solutions and elevated temperature on the dynamic modulus of elasticity, compressive strength and flexural strength were not measured by these authors.

Generally, this chapter was written to firstly recognise the most significant properties of GPC which should be examined in this research work before establishing the limits within which waste materials such as fly-ash and bottom-ash can be used as a cement replacement material in concrete. The second reason was to attain a good understanding of the factors which affect numerous properties of GPC so that the experimental programme can be designed to study the comparative influence of waste materials on properties of GPC. Based on the above literature review, it is concluded that materials such as fly-ash and bottom-ash from different industries can be used in varying proportions for partial replacement of concrete ingredients. Moreover, detail investigation on the different properties of GPC including leach-ability, freeze-thaw resistance and elastic modulus, etc. is required for utilization of these waste materials in the construction sector.

# Chapter 3   An Overview on General Methodology

This Ph.D dissertation includes the following major methodology to address the research knowledge gaps:

**Phase (I)** Fly-ash is one of the most used materials for making GPC. However, bottom-ash with similar chemical components as fly-ash can also be used as source material, since the use of this waste material is almost ignored in the construction of GPC. Moreover, although, sodium-based GPC made by either fly-ash or slag is not new, there are limited studies on potassium-based GPC made by the combination of bottom-ash and fly-ash. So, first, this study focuses on the production of potassium-based GPC using fly-ash and bottom-ash and investigate the durability of this type of concrete under various conditions such as freeze-thaw cycles. To develop GPC, different types of the mixture were produced at the Civil Engineering Materials lab of the University of Victoria. The durability of structures is greatly influenced by the curing of GPC. The duration of curing also affects other properties of GPC such as water absorption, porosity and permeability. Thus, since it was found that GPC is a sensitive material to moisture and temperature, attempts have been made to cure the 54 GPC samples using two different accelerating curing methods (dry curing and steam curing) to achieve high strength at an early age. According to the results, the highest compressive strength has achieved by steam cured GPC samples made by 50% bottom-ash and 50% fly-ash. Furthermore, the elastic modulus of GPC is a significant property because it is vital for controlling deformation. There are numerous studies predicting the "static" elastic modulus of GPC made by various waste materials such as fly-ash and slag. However, this research aims to make a relationship between "dynamic" elastic modulus and compressive strength of 12 GPC samples made by both fly-ash and bottom-ash using NDTs including resonant frequency test, since the authors could not find any predictive model for this type of concrete.

**Phase (II)** Micro-structural study on concrete is an exclusive method to investigate the morphological characteristics of concrete. However, there is no study on microstructure of GPC in a frozen condition because most of the scanning electron microscopic devices cannot maintain the low temperature required to take an image from a frozen GPC specimen. So, a novel study was performed on the microstructure of GPC in a frozen condition using 4-D Low Temperature Scanning Electron Microscopic (4-D LTSEM) device. In this regard, tiny GPC samples were frozen at various temperatures ranging from 0 °C to -180 °C. This work was mainly performed to investigate the applicability of practice guideline/standard for GPC. The obtained results show that the rate of cryo formation in air voids of GPC is slow from 0 °C to -18 °C, showing adequate resistance of GPC when exposed to frozen climates.

**Phase (III)** It is a well-established fact that freeze-thaw damage is one of the major types of concrete deterioration. Due to the lack of knowledge on freeze-thaw resistance of GPC made by fly-ash and bottom-ash, this current work proposes the freeze-thaw resistance of GPC conducted by the accelerated test method. Thus, attempts have been made to use Non-Destructive Tests (NDTs) (such as resonant frequency test) to determine the freeze-thaw durability of 12 bottom-ash based GPC and 12 fly-ash based GPC from their mass loss and dynamic elastic modulus. The goal of this phase of the current study is to derive a model for predicting the life-span of concretes exposed to freeze-thaw cycles. Moreover, one of the major concerns in the production of GPC is the increase in risk of contamination of final products due to the use of waste materials and chemical activators. Many environmental and laboratory studies carried out on conventional and fly-ash based/sodium-based GPC. However, the author could not find any other work about the leaching of heavy metals of bottom-ash based GPC. So, in this regard, Toxicity Leaching Characteristic Procedure (TCLP) test was used to trace the heavy metals of bottom-ash based GPC. The results indicate that this type of concrete has a high degree of immobilization of heavy metals.

**Phase (IV)** Steel reinforcement corrosion has been widely stated in the literatures over the last decades. It is one of the major durability problems because steel corrosion in GPC causes cracking, reduction of bond strength, and loss of life span. So, it emerges that the new type of concrete (GPC) must be experimentally studied concerning the corrosion behaviour of steel in GPC, to make reliable service life prediction. In this study, 3 GPC and 3 PCC specimens were tested for corrosion rate. Two NDTs including Half-cell potential and linear polarization resistance were used to measure the corrosion rate of steel reinforced GPC and

PCC. The result indicated that GPC has a higher resistance to chloride-induced corrosion; with low corrosion rate and lower mass loss percentage, compared to PCC.

**Phase (V)** It is well-known that most of the experiments on GPC are performed at the laboratory and controlled conditions. That is why, in comparison to previous studies, since there is no large-scale work (in real environmental conditions) on bottom-ash based GPC, paver blocks were selected as an application. The 150 GPC paver blocks were cast at the laboratory and were placed at a parking lot to study the effects of real environmental conditions on the durability of GPC using NDT. For comparison, 240 Portland Cement Concrete (PCC) paver blocks with almost the same mechanical properties as GPC were commercially purchased to have a better idea about the durability of these types of concrete in the same condition. This part of the study is divided into two major phases including freeze-thaw (wet-dry) durability and leach-ability of toxic metals. In this regard, a couple of NDTs including ultrasonic pulse velocity, HACH strips, Schmidt hammer, etc. were used to evaluate the durability of GPC and PCC paver blocks. Ultimately, the results of NDTs were used to derive a predictive model for freeze-thaw damage using dynamic elastic modulus.

# Chapter 4    Studying and Predicting the Elastic Modulus Characteristic of GPC

## 4.1    Introduction

The objective of this study is to produce cement-less concrete using waste materials including fly-ash and bottom-ash. GPC as an alternative to PCC is selected in this research to reduce the environmental impact of cement production/$CO_2$ emission. Even though, there are numerous studies reported on GPC made by fly-ash, limited studies report on GPC made by 50% fly-ash and 50% bottom-ash.

Moreover, it is found by the current researchers that GPC is a sensitive material to moisture and temperature. Heretofore, the main limitation of producing GPC is related to the low compressive strength attained under ambient conditions [58]. The mechanical properties of GPC are found to improve after being heat-cured [59,60]. Several efforts [32,61,62] were concluded for determining the influence of curing conditions on GPC made by different precursors including fly-ash. However, since, there is limited work on curing regime of GPC made by the combination of fly-ash and bottom-ash, three methods of curing (ambient, steam and dry) were used to enhance the geopolymerization process. In these methods, GPC samples were exposed to various temperatures including ambient, 30 °C, 45 °C, 60 °C, and 80 °C for 24 hours, followed by 28 days of ambient curing.

Elastic modulus is an important property of the material that indicates its resistance to being deformed elastically [25]. Many empirical models for prediction of elastic modulus of fly-ash based GPC have been proposed. However, in this study attempts have been made to establish a comparison between experimental results of elastic modulus and proposed predictive models to find a model that fits for bottom-ash based GPC.

4.2 Curing regime and elastic modulus of GPC
The following manuscript has been published in the journal of "Buildings". It should be noted
that this chapter is an edited version of the published manuscript.

## 4.3    Introduction

One of the main environmental issues associated with the production of cement is Green House Gas (GHG) emissions to the atmosphere, as well as the consumption of natural resources. It is estimated that one ton of cement production needs about 1.5 tons of natural resources and emits about one ton of $CO_2$ into the air [63-65]. On the other hand, about 112 million tons of fly-ash as waste material is produced annually in the world [66]. So, the use of fly-ash in the concrete mixture can produce positive environmental, economic, and product benefits [67]. However, about 80% of the unburned material or fly-ash is entrained in the flue gas when pulverized coal is burned in a dry condition. The remaining 20% of the ash is dry bottom-ash that has similar chemical properties to fly-ash, consisting of silica and alumina. In the United States in 2017, 38.2 million tons of fly-ash and 9.7 million tons of bottom-ash were generated [68]. However, the use of bottom-ash is relatively limited, thus dealing with this industrial material is posing to be one of the most significant challenges in recent years. This is why a combination of bottom-ash and fly-ash was used to develop a cement-less concrete mix by the authors in this study. To the authors' knowledge, limited studies have been done that deal with this combination [69-71].

GPC is a type of cement-less concrete that is produced using alumino-silicate material such as fly-ash and has the potential to decrease the significant carbon footprint associated with the production of cement. GPC is an alumino-silicate polymer (inorganic) which is made using geological origin materials or industrial materials including fly-ash, slag, metakaolin, etc. The polymerization process involves a substantial chemical reaction of alumino-silicate species under alkaline conditions, resulting in a three-dimensional polymeric chain, including (i) dissolution of silicon and aluminum atoms through the hydroxide ions reaction, (ii) transportation, orientation and condensation of ions into monomers and (iii) polycondensation/polymerization of monomers into polymeric chains [72]. Basically, the Life Cycle Assessment (LCA) method is used to calculate the range of $CO_2$ emission of materials, considering all processes such as raw material (natural sources), transportation, manufacturing and use. According to the studies by McLellan et al. [73] and Habert et al. [74], due to the mining, processing, and transport of raw materials, the range of $CO_2$ emission for GPC is 26%–45% lower compared to cement-based concrete. Numerous researchers have studied the durability of Na-based GPC [8,75]. However, there are limited studies conducted on the development of mix design, durability and elastic modulus determination of K-based GPC made using a combination of bottom-ash and fly-ash [25].

The mechanical properties of GPC are influenced by various factors including raw materials, the activator type and molarity [76,77]. Ryu et al. [42] and Khater et al. [78] reported that curing condition also plays a vital role in dissolution and geopolymerization of the aluminum-silicate gel which results in high early strength gain. Therefore, the need for sufficient curing of concrete cannot be overemphasized. According to reported literatures [79,80], GPC requires heat treatment in order to attain similar or higher compressive strength in comparison with PCC. Yang et al. [81] compared the emission of $CO_2$ from both GPC and PCC. The results showed that the total $CO_2$ emission from heat curing (steam curing at 85 °C for a duration of 24 hours) is evaluated at about 38.5 kg/FU, which is quite energy intensive. However, this energy consumption in GPC is reduced in the transportation phase, where the total $CO_2$ emissions of GPC can be 45% less when compared to its counterpart PCC. In this study two methods of curing (steam and dry) were used to investigate their influence on compressive strength for GPC made using 50% fly-ash and 50% bottom-ash. This presented mix proportion/ratio used in this study was chosen after considering several other proportions in separate studies [82-84] previously completed by the authors and their associates. Sajedi et al. [85] compared the compressive strength of 24 mixes cured with various curing methods including room temperature, in water without heating, room temperature after heating 60 °C for duration of 20 hours and in water after heating 60 °C for duration of 20 hours. According to the results, due to relatively high consistency of slag when exposed to unheated water, the slag-based mortars showed higher compressive strength (80 MPa) at 90 days compared to other mixes. Parghi et al. [86] investigated the effect of various curing temperatures and curing duration (ambient, moisture curing and submergence curing) on engineering properties of polymer-modified mortars. In accordance to results, polymer-based mortars (polymer/cement ratio up to 15%) cured for seven days in water submergence and for 21 days in ambient temperature gained higher durability compared to other types of mix composition and curing regime. Parghi et al. [86] attributed this phenomenon to a greater bond strength of both cement particles and polymer films. Heah et al. [87] microstructurally studied the influence of a curing regime on kaolin-based GPC. The samples were cured at temperatures of ambient, 40 °C, 60 °C, 80 °C and 100 °C for one day, and up to three days. The SEM results showed that higher geopolymeric gel and denser matrices was developed for kaolin-based GPC cured at 60 °C up to three days.

Basically, hydration is a chemical reaction in which the major compounds in cement create chemical bonds with water molecules and become hydrates or hydration products. The majority reaction product formed in cement hydration is a C-S-H gel to which the material mostly owes its durability. The secondary products include portlandite, ettringite and calcium monosulfoaluminate. Moreover, to achieve optimum compressive strength at ambient temperature, it is reported that calcium-based materials such as calcium oxide (CaO) contributes to form hydrated gel including calcium-silicate-hydrates (C-S-H) along with the alumino-silicate gel, where CaO content leads to durable GPC [88]. However, in this study attempts have been made to produce GPC using bottom-ash and fly-ash using heat treatment.

Moreover, establishing a relationship between various mechanical properties of GPC is important for future applications of GPC. The modulus of elasticity is a vital mechanical property of concrete including GPC when compression stress is applied. Modulus of elasticity in concrete refers to the deformation resistance a concrete structure has to loading. That is, elastic modulus evaluates the stiffness of the concrete. This property is used for design of various structural applications such as high-rise buildings and bridge piers where the stiffness of the concrete is of great importance. Noushini et al. [89] produced class F fly-ash based GPC samples to measure the effect of heat curing on their mechanical properties including compressive strength and elastic modulus. The GPC samples were cured at temperatures of ambient, 60 °C, 75 °C and 90 °C for durations of 8, 12, 18 and 24 hours. According to the results, the compressive strength of $62.3 \pm 0.2$ MPa and elastic modulus of 25.9 GPa was achieved with the increase in the curing temperature up to 75 °C and curing duration up to 24 hours. Yang et al. [36] made a correlation between compressive strength and elastic modulus of various mixes. Yang et al. [36] observed that with similar compressive strength (about 30 MPa), heat-cured GPC made using only fly-ash had lower resonant frequency than PCC, which meant a lower modulus of elasticity. However, there are limited studies on the modulus of elasticity of GPC made using bottom-ash [25]. The authors also could not find any work that predicts and compares these values with PCC.

## 4.4 Research contribution

In meso-scale level, this study aimed at developing a new type of concrete made using a combination of waste materials including fly-ash and bottom-ash. Even though extensive research has been performed on Na-based GPC made using different precursors [90-94], limited studies are reported in the literature on K-based GPC especially when made using a combination of 50% fly-ash and 50% bottom-ash.

GPC as a material is known to be sensitive to moisture and temperature [21,77]. Moreover, heat treatment has a significant impact on the microstructure and strength development of GPC. Hence, the geopolymerization process is normally dependent on methods of curing. This study is also targeted at experimentally investigating the influence of the heat curing regime on the compressive strength of GPC. Two methods of curing (steam and dry) were used in order to gain similar or higher compressive strength in comparison with PCC. Moreover, it is reported [95,96] that higher curing temperature and curing time result in higher ultimate strength of GPC. In this study, attempts have been made to determine optimum curing temperature and duration for K-based GPC made using fly-ash and bottom-ash.

The elastic modulus is also one of the key mechanical properties of GPC that indicates its ability to deform elastically. A concrete structure needs to have higher compressive strength with a higher elastic modulus to allow carrying of high load values. Otherwise, lower elastic modulus limits the practical applications since the deformations experienced are very high, causing failure of serviceability limits. Since the authors have not found any other work reported in the literature on correlating elastic modulus and compressive strength of K-based GPC made using bottom-ash and fly-ash, in this study attempts have been made to experimentally find a better predictive model for GPC using the proposed empirical formula by other researchers.

## 4.5   Experimental program

### 4.5.1   Fly-ash and bottom-ash

Class F fly-ash and bottom-ash used in this study were obtained from pulverized coal combustion from Lafarge Canada Inc. The chemical compositions of bottom-ash and fly-ash (shown in Table 1) were determined by X-Ray Diffraction (XRD) technique. The major components in fly-ash and bottom-ash were silicon oxide, aluminum oxide and iron oxide.

**Table 1. Chemical composition of fly-ash and bottom-ash. Source: Lafarge Canada Inc.**

| Chemical Compounds | Fly-Ash (%) | Bottom-Ash (%) |
|---|---|---|
| $SiO_2$ | 47.1 | 60.11 |
| $Al_2O_3$ | 17.4 | 14.35 |
| $Fe_2O_3$ | 5.7 | 5.92 |
| $CaO$ | 14 | 10.40 |
| $MgO$ | 5.4 | 4.49 |
| $SO_3$ | 0.8 | 0.10 |
| LOI | 0.19 | 0.00 |
| $Na_2O$ | N/A | 2.232 |
| $K_2O$ | N/A | 1.766 |
| $TiO_2$ | N/A | 0.892 |
| $P_2O_5$ | N/A | 0.200 |
| $Mn_2O_3$ | N/A | 0.093 |

Table 2 shows the physical properties of fly-ash particles measured at the Lafarge Concrete lab. The measured properties are compared with the requirement of ASTM C618 [97]. The physical properties of bottom-ash particles are not evaluated by Lafarge because bottom-ash is not characteristically used in the construction industry. In general, bottom-ash particles are much coarser than fly-ash particles. This is why bottom-ash particles were sieved (#1.18) to eliminate large particles and to increase the surface area of particles to attain optimum strength.

**Table 2. Fly-ash physical properties**

|  | Fly-Ash | ASTM C618 [97] |
|---|---|---|
| **Fineness retained on 45 μm (No. 325 sieve)** | 17.3% | <34% (complies) |
| **Strength activity index with Portland cement% of control at 28 days** | 99% | >75% (complies) |
| **Water requirement, percent of control** | 100% | <105% (complies) |
| **Autoclave expansion** | 0.04% | <0.8% (complies) |
| **Density** | 2.65 Mg/m$^3$ | <5% (complies) |

Table 3 shows other properties of fly-ash and bottom-ash obtained from the Material Safety Data Sheets (MSDS) of Lafarge Canada Inc.

**Table 3. Other properties of both fly-ash and bottom-ash**

| Material | Appearance | Odor | pH | Boiling Point | Specific Gravity | Solubility |
|---|---|---|---|---|---|---|
| **Fly-ash** | Gray/Black or Brown/Tan (Powder) | odorless | 4–12 | >1000 °C | 2.0–2.9 (water = 1) | Water: <5% (slightly) |
| **Bottom-ash** | Gray/Black or Brown/Tan (Powder) | odorless | 4–12 | >1000 °C | 2.0–2.9 (water = 1) | Water: <5% (slightly) |

Figure 1 presents the images of both fly-ash and bottom-ash taken by a Hitachi S-4800 SEM at the University of Victoria. When using the SEM, accelerating voltage and magnification of 15 KV and 2200x, respectively, were used. Noticeably, diverse shape of bottom-ash particles shown in Figure 1(i) is visible, ranging from almost impeccably spherical to extremely irregular (rough formed) shaped particles. It can be seen in Figure 1(ii) that fly-ash particles are glassy and spherical in shape. Moreover, the particle size investigation of bottom-ash and fly–ash performed by the authors showed that the mean size of rounded shaped bottom-ash and fly-ash is 58.53 μm and 10 μm, respectively [98]. The average height and width of the rough formed bottom-ash is 22.59 μm and 12.78 μm, respectively.

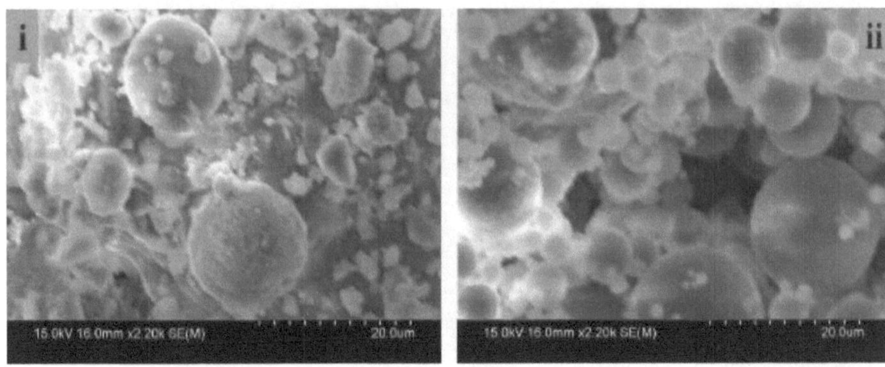

**Figure 1. SEM images of (i) bottom-ash (ii) fly-ash.**

### 4.5.2 Alkaline liquid and aggregates

A combination of Potassium Silicate ($K_2SiO_3$) solution and Potassium Hydroxide (KOH) solution was chosen as the alkaline liquid. KOH solution was prepared using industrial grade KOH flakes obtained from Sigma-Aldrich Private Ltd., Canada. The KOH solution was prepared by dissolving the flakes in water. The mass of KOH solids in a solution varied depending on the concentration of the solution expressed in terms of molarity (M). $K_2SiO_3$ powder (AgSil 16) obtained from PQ Corporation (USA) was used. Based on the MSDS of the product, the chemical compositions of $K_2SiO_3$ powder were $K_2O = 32.4\%$, $SiO_2 = 52.8\%$ and water = 14.8% by mass.

GPC was produced with sand and coarse aggregates (with a maximum size of aggregates 12.5 mm) from a quarry in British Columbia, Canada with a relative dry density of 2.67 and 2.71, respectively. The water absorption ratio of sand and coarse aggregates was 0.79% and 0.69%, respectively, in accordance with ASTM C127 [99]. The particle size distribution of coarse and fine aggregates was measured, shown in Figure 2, in accordance with ASTM C33 [100]. Fineness moduli of the coarse and fine aggregates were measured as 6.85 and 3.54, respectively.

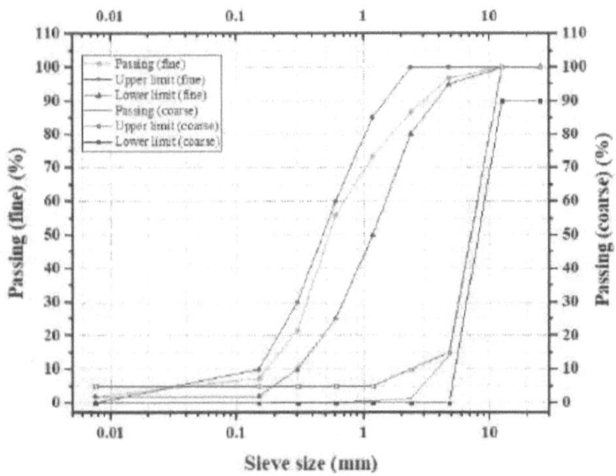


**Figure 2. Aggregates' size distribution**

### 4.5.3 Mix proportions

The goal of this study was to compare the performance of the developed GPC with a comparable PCC mix. Since the binders in both PCC and GPC are completely different, their contents had to be varied to achieve a comparable target strength of 35 MPa. The mix proportion of GPC was derived after initial trial experiments on mortars performed in the Facilities for Innovative Materials and Infrastructure Monitoring (FIMIM) by the authors and their associates [82-84].

Table 4 shows the mix proportions of GPC and PCC used in this study. As shown in Table 4, the mass ratio of fly-ash to bottom-ash is 50:50 and the concentration of 12 M was derived after initial testing on GPC. The ratio of alkaline liquid to fly-ash and bottom-ash was selected as 0.21 approximately.

**Table 4. Mix design**

| | GPC | PCC |
|---|---|---|
| Material | Content (kg/m$^3$) | Content (kg/m$^3$) |
| Fly-ash | 194 | — |
| Bottom-ash | 194 | |
| Coarse aggregates | 1170 | 1120 |
| Sand | 630 | 820 |
| KOH (12 M) | 85.16 | — |
| K$_2$SiO$_3$ | 125.74 | |
| Cement (PC) | | 340 |
| Water | 38.71 | 181 |

Ordinary Portland cement-Type I in accordance with ASTM C150 [101] (referred to as type General Use (GU) in CSA A23.1-14 [102]) was used in the production of PCC. As can be seen in Table 4, the water/cement ratio of approximately 0.53 was selected with a target strength of approximately 35 MPa. The target strength of 35 MPa for both types of concrete was selected as this strength is typically prescribed for several common applications in practice.

## 4.6    Method of casting, curing and testing

### 4.6.1    Specimen preparation

Typically, the concrete paste design is critical and depends on the application where it is going to be used. In this study, curing temperature, duration and consequently its effect on the compressive strength of GPC were chosen as the key focus.

In this study, GPC and PCC cylindrical samples were cast and cured as detailed in ASTM C192 [103]. For the manufacturing of GPC, first of all, a combination of $K_2SiO_3$ and KOH (alkaline solution) was prepared 24 hours before mixing. Later, sand and coarse aggregates were placed into a mixer, and the mixing procedure was initiated. After 30 seconds, both ashes were added to the mix after stopping the mixer. Then, the alkaline solution was added without pausing the mixer. After 30 more seconds, part of extra water (if needed) was added (and noted) to the mixer to achieve the desired workability. After all ingredients were in the mixer, the materials were mixed for 3 minutes followed by a 3-min rest period, followed by another 2 minutes of final mixing. After 2–3 minutes of rest, cylindrical samples ($\Phi 100 \times$ 200 mm) were prepared to measure the compressive strength of mixes cured at various temperatures in both dry and steam curing conditions. After 30 seconds of vibration of samples on a table vibrator, each cylinder was covered with a plastic sheet to prevent excessive evaporation and kept overnight in the laboratory at ambient temperature.

For PCC, the sequence of mixing was as follows: coarse aggregates and sand were mixed in dry form for 1 minute. Then, cement was added to the mixer and mixed for 2 minutes. After proper mixing of dry materials, water was gradually added while the mixer was in motion. After adding water, concrete was mixed for 2 minutes in the mixer, then rested for 1 minute, followed by mixing again for 2 minutes.

### 4.6.2    Curing of samples

Accelerated curing is any method by which high early age strength is achieved in GPC. In this paper, dry and steam curing methods were used to achieve higher compressive strength. Hence, for the dry curing method, GPC specimens were kept in an oven at temperatures of 30 °C, 45 °C, 60 °C, and 80 °C for a period of 24 hours and kept at ambient temperature for 28 days. For the steam curing method, samples were kept in a bucket surrounded by water.

The bucket was wrapped up tightly to prevent excessive evaporation during the curing process and the bucket was then kept into an oven at temperatures of 30 °C, 45 °C, 60 °C, and 80 °C. Lastly, samples were removed from the oven after 24 hours, unwrapped and kept at ambient temperature for 28 days. It is well established that a higher temperature for a longer period of time (up to 72 hours) can result in higher strength gain in GPC [95,96]. However, the aim of this study was to expose the specimens to heat curing only for 1 day to reduce the amount of energy utilized for curing. Another goal was to determine the feasibility of using curing temperatures lower than those reported in the literature [104].

The PCC samples were cured by demolding and storing in a water tank at 23 ± 2 °C for 28 days.

## 4.7 Test methods
### 4.7.1 Compressive strength

A total of 54 GPC and 6 PCC cylindrical samples ($\Phi100 \times 200$ mm) were tested at an age of 28 days using a Forney compression test machine in accordance with ASTM C39 [105]. The loading rate was between 0.15–0.3 MPa/s. According to ASTM C39 [105], the results of properly conducted tests on two trial batches made in the same laboratory should not differ by more than 3.95 MPa. The single-operator standard deviation for compressive strength of trial batches has been found to be 1.39 MPa.

### 4.7.2 Dynamic modulus of elasticity

To evaluate the dynamic modulus of elasticity of GPC and PCC specimens, an additional 24 samples were cast and cured for 7 and 28 days. The RFT was conducted in accordance with ASTM C215 [39]. The ASTM C215 [39] is often considered for elastic modulus calculation of conventional concrete. However, it should be noted that the elastic modulus of the final element is directly linked to mechanical strength/compressive strength development of the material. Moreover, GPC has almost similar composition as conventional concrete except for the binder. Elastic modulus is known to be highly dependent on the aggregate type and quality. So, procedures in ASTM C215 [39] should also be applicable for determination of the elastic modulus of GPC. Longitudinal impact RFT was performed on cylindrical specimens of 100 mm diameter and 200 mm height as seen in Figure 3.

**Figure 3. Fundamental longitudinal RFT**

The RFT device includes a PCB Integrated Circuit Piezoelectric (ICP) accelerometer (with a pickup sensitivity of 102.2 mV/g (10.2 mV/ms$^{-2}$) and frequency range of 0.3–15 kHz) that is attached to the concrete sample surface using adhesive wax and a 4 channel NI USB6009 DAQ Analog Digital Converter (ADC) is used for collecting the input compression wave voltage. A standard ball tip hammer weighing $110 \pm 2$ g with a tip diameter of 10 mm is used to strike the sample surface. A total of 72 readings on 24 samples were taken and outlier were omitted in the average values. National Instrument's LabVIEW 2014 (NI LabVIEW) was used to acquire and analyze the data. The program was developed for this study considering ASTM C215 [39] for acquisition requirements. Table 5 indicates the properties of the accelerometer used in this study.

**Table 5. Accelerometer specification**

| Type | PCB-ICP |
|---|---|
| Model | 352C33 |
| Height | 15.7 mm |
| Weight | 5.8 g |
| Sensitivity | 10.2 mV (m/s$^2$) |
| Frequency Range | 0.3–15,000 Hz |
| Non-Linearity | <1% |
| Transverse Sensitivity | <5% |
| Temperature Range | −54 to 93 °C |

The dynamic elastic modulus was calculated using Equation (2) in accordance with ASTM C215 [39]:

$$E = 5.093 \ (L/d^2) \ Mn^2 \qquad (2)$$

where:

    E = dynamic elastic modulus (Pa)

L = length of concrete cylinder (m)
d = diameter of concrete cylinder (m)
M = mass of concrete cylinder (kg)
n = fundamental longitudinal frequency (Hz)

## 4.8 Modulus of elasticity predictive models

Various modulus of elasticity predictive models (discussed below) of different GPC mixtures were compared with the modulus of elasticity of K-based GPC obtained from the experimental results in this study. This was with a goal to identify the most suited model for the material developed in this study.

Hardjito et al. [106] established Equation (3) for calculation of elastic modulus of heat-cured fly-ash based GPC.

$$E_c = 2707 \sqrt{f_{cm}} + 5300 \tag{3}$$

E = Modulus of elasticity (MPa)

$f_c = f' = f_m$ = Compressive strength (MPa)

$\rho$ = Density (kg/m$^3$)

The aforementioned variables and units are used henceforth.

Hardjito and Rangan [107] made a comparison between experimental values of elastic modulus of fly-ash based GPC, the empirical formula proposed by AS 3600 [56] and ACI Committee 363 [108]. The elastic modulus of specimens was measured in accordance with the Australian Standard AS 1012.17 [109]. According to the results, experimental values were lower than the proposed model due to the granite-type coarse aggregates used in the experimental results. The formula proposed by AS 3600 [56] was not used in the current study because it is established for concrete that has compressive strength greater than 40 MPa. Hardjito and Rangan [107] recommended that the ACI 363 [108] model (Equation (4)) is better for calculating the modulus of elasticity.

$$E_c = 3320 \sqrt{f'_c} + 6900 \tag{4}$$

Tempest et al. [110] proposed Equation (5) to calculate the elastic modulus of fly-ash based GPC. The modulus of elasticity of samples was evaluated using the procedure given in ASTM C469 [111]. The results showed that the elastic modulus increases as the compressive strength of GPC increases. It was also found that experimental values have the same trend (slightly higher) as the proposed formula.

$$E_c = 3421\sqrt{f'_c} \tag{5}$$

Prachasaree et al. [112] measured the static modulus of elasticity of specimens in accordance with ASTM C469 [111]. They used a predictive model proposed by Hardjito et al. [106], ACI 318 [57] and ACI 363 [108] to calculate the elastic modulus of GPC. However, it is stated that all the proposed formulas do not fit the experimental results. Therefore, they proposed Equation (6) for calculation of elastic modulus of fly-ash based GPC.

$$E = 840 - 886\sqrt{fc} + 647f_c \tag{6}$$

Thomas and Peethamparan [113] proposed two equations for the prediction of elastic modulus from compressive strength using ASTM C469 [111]. The difference between these two equations was not clearly stated in their study. The proposed equation by ACI 318 [57] was compared with Equations (7) and (8). The results showed that both Equations (7) and (8) vary slightly from ACI 318 [57] in terms of predicted values and goodness of fit. However, authors mentioned that due to variation in the data, additional investigation is needed to find a better fitting formula.

$$E = 2900f_c^{3/5} \tag{7}$$

$$E = 4400\sqrt{f_c} \tag{8}$$

Wardhono et al. [114] compared the experimental results of modulus of elasticity of fly-ash based GPC with proposed equations by Ng and Foster [115], AS 3600 [56] and Diaz-Loya et al. [116]. It is reported that AS 3600's model with higher $R^2$ (98.13%) showed a closest and similar trend to experimental values. However, only five specimens were used to determine the modulus of elasticity of fly-ash based GPC at different ages including 28, 90, 180, 360, 540 days. In this study, Equation (9) was established for the determination of elastic modulus.

$$E_c = f_c^{1.6412} \, 49.968 \tag{9}$$

Gunasekara et al. [117] made a correlation between elastic modulus and compressive strength of fly-ash based GPC using equations proposed by Hardjito and Rangan [107], Diaz-Loya [116], Olivia et al. [45] and Wardhono et al. [114]. They also compared the determined formula with AS 3600 [56]. The static modulus of elasticity of samples was measured in accordance with AS 1012.17 [109]. It is reported that AS 3600 [56] overestimated the experimental results of GPC. Equation (10) was developed for the determination of elastic modulus:

$$E_c = \rho^{1.5}\, 0.024\sqrt{f_c} \qquad\qquad (10)$$

In this study, all the predictive models reported above have been compared to test results presented later in this manuscript.

## 4.9    Results and discussion

### 4.9.1    Workability and density

The slump test was performed immediately after preparing the mixes to measure the workability of fresh paste in accordance with ASTM C143 [118]. In this study, the slump value of K-based GPC and PCC was 185 mm and 210 mm, respectively. It is reported that GPC and PCC are considered to be highly workable when GPC and PCC gain a slump value over 90 mm and 180 mm, respectively [119,120]. These two types of concrete with high workability are very easy to mix, transport, place and compact in structures. K-based GPC showed lower workability compared to its counterpart PCC because of the existence of silicates in GPC which increases the viscosity characteristic of the paste and makes GPC more cohesive and stickler than PCC [121,122].

The average dry density of three PCC and GPC samples at seven days and 28 days was measured. The results showed that the average density of GPC increased from 2430 kg/m$^3$ at seven days to 2437 kg/m$^3$ at 28 days with an overall increase of 0.28%, while the average density of PCC increased from 2453 kg/m$^3$ at 7 days to 2462 kg/m$^3$ at 28 days with an overall increase of 0.36%. These results can be attributed to the microstructure of bottom-ash particles, where large size and irregularly-shaped particles of bottom-ash caused GPC to be ever so slightly less dense and nonhomogeneous than PCC [25]. Moreover, this finding is in good agreement with the study of Chindaprasirt et al. [69] and Haq et al. [71] where bottom-ash based GPC showed lower dry density due to the excessive evaporation of residual liquid such as water and alkaline solution over the curing process.

### 4.9.2    Effect of heat treatment on compressive strength of GPC samples

As mentioned earlier, in this study, the compressive strength of hardened concrete was selected as the performance criteria. Generally, longer curing time enhances the polymerization process resulting in higher compressive strength. In this study, five different curing temperatures were used, i.e., ambient, 30 °C, 45 °C, 60 °C, 80 °C. Figure 4 presents the average compressive strength of steam and dry-cured samples. The results are reported as an average of six samples along with error bars. Figure 4 indicates that the compressive strength of steam-cured samples increased about 3.5 times when the temperature went from ambient temperature (~10 °C) to 80 °C. In comparison, the compressive strength increased approximately 2.3 times when specimens were dry-cured. Figure 4 also shows that the temperature of 80 °C during the steam curing method resulted in higher compressive strength. Based on the results of compressive strength, steam curing at the temperature of 80 °C was chosen in this study as the preferred method of curing.

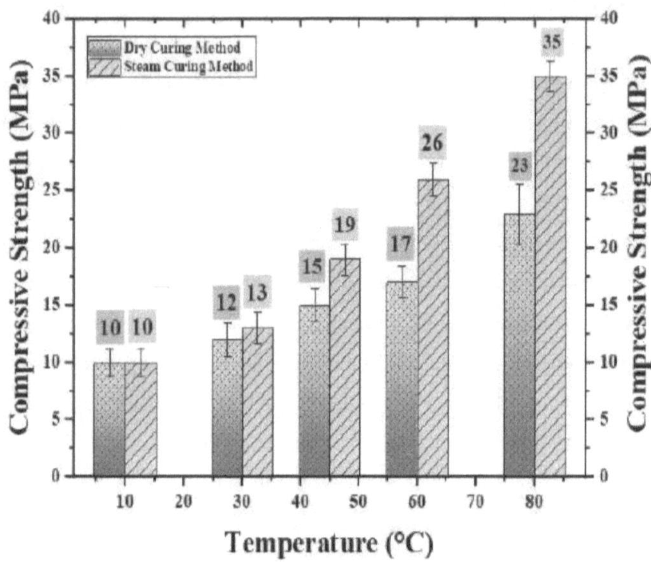


**Figure 4. Average compressive strength of GPC sample**

According to other studies on the microstructure of GPC [123,124], the steam curing method improves the dissolution rate of chemical species, such as Silicon Dioxide ($SiO_2$) and Aluminum Oxide ($Al_2O_3$), from paste where the rate of geopolymerization increases. This finding can be attributed to the full and uniform internal curing of specimens. Abdollahnejad et al. [125] microstructurally investigated the effect of thermal curing on fly-ash based GPC. The results showed that thermal treatment causes denser paste and less cracks due to the higher degree of reaction and higher adhesion between the amorphous gel and the particles. It should be noted that an increase in compressive strength is typically not expected to be achieved beyond a temperature of 80 °C [54], hence this was established as the upper bound for curing temperature in this study.

Figure 4 also reveals that at the same curing temperature, steam-cured samples yielded a higher strength than that of dry-cured samples. This can be attributed to the existence of the adequate amount of water during the geopolymerization process, while, in the dry curing method, an excessive amount of water evaporated quickly from samples and there was not enough water in the samples for complete geopolymer formation [126]. Generally, the geopolymerization process is accelerated at the higher temperature [127]. In this study, ambient-cured GPC samples did not develop adequate compressive strength at an age of 28 days because GPC paste reacts slowly and takes considerably long time to set at ambient/low temperature compared to heat-cured samples [128]. However, it is suggested to add calcium-based materials (such as cement and slag) in GPC mixture to improve engineering properties of ambient-cured GPC since it is more practical in-situ to cure samples at ambient temperature [88,129].

Provis et al. [130] reported that the Na-based solution is mostly used because of its availability, low cost, and high reactivity. However, the K-based solution is considered for high temperature applications [131,132]. Hounsi et al. [133] reported that Na-based GPC has lower compressive strength value than K-based GPC at the same alkaline concentration of the current study (12 M) due to the reduction of Si/Na ratio at high concentration of Sodium Hydroxide (NaOH), where NaOH slows the polycondensation process and damages the mechanical properties of Na-based GPC.

Average compressive strength of 31 MPa and 33 MPa for PCC was measured at an age of 7 days and 28 days, respectively. This was most comparable to the strength of steam-cured GPC samples at 80 °C.

It is reported that the compressive strength of 17 MPa is appropriate for residential buildings and compressive strength of 28 MPa and higher is proper for commercial buildings [134]. Greater compressive strength up to and beyond 70 MPa is stated for other applications such as dams [134]. Therefore, selected compressive strength/temperature (35 MPa/80 °C) in this study provides and meets a wide range of requirements for applications including residential and commercial. However, it should be mentioned that GPC samples can be either dry-cured or steam-cured at 60 °C for some applications and it may not be necessary to cure at a high temperature such as 80 °C.

Generally, IS 456 [135] has designated the concrete mixes into a number of grades as M10, M15, M20, M25, M30, M35 and M40. According to IS 456 [135], generally, M10 grade of concrete can be used for leveling course, and bedding for footings. Moreover, the target strength of the current study (M35) can be used as structural concrete including commercial structures, external walls and slabs, as well as for structural piling.

It is well-known that only ambient temperature curing is a practical method in the construction of concrete buildings or other structures. Hence, according to Nath et al. [136], using calcium-based material such as cement enhances setting time, workability and durability of GPC cured at ambient temperature. The microstructural investigation of high-calcium-based GPC indicated that both sodium aluminate silicate hydrate (N-A-S-H) and calcium aluminate silicate hydrate (C-A-S-H) gel are formed in high-calcium-based GPC [125]. This phenomenon is reported by other studies [125,137,138], where silica content, calcium content and aging conditions affect the phase transition behavior of fly-ash based GPC. Abdollahnejad et al. [125] reported that the samples with higher molar ratio of Na/Ca achieved higher rates compared to other samples with molar ratio of Si/Al and Ca/Si in terms of gel formation at the ambient temperature (25 °C and 30% relative humidity). Thus, the replacement of calcium by sodium causes ionic exchange mechanisms, where sodium species are replaced by calcium ions ($Ca^{2+}$) in C-A-S-H gel, and (C,N)-A-S-H gel is formed [125]. This replacement causes a denser paste and decreases crack creation. Abdollahnejad et al. [138] replaced 10%, 20%, and 30% ceramic wastes with slag to produce GPC, containing coarse porcelain ceramic waste as aggregate. Two methods of the curing regime include sealing with plastic and thermal curing conditions for 3 hours at 60 °C to cure the samples. According to the results, lower amounts of calcium content caused a reduction in compressive strength of GPC samples.

### 4.9.3 Longitudinal resonant frequency

The average resonant frequency of six GPC and six PCC specimens was calculated. Figure 5 shows the values of resonant frequency versus compressive strength at an age of 7 and 28 days, respectively. It can be seen in Figure 5 that the resonant frequency of both types of concrete increased gradually as the age increased due to the continuous geopolymerization/hydration process. According to the results, after 28 days of curing, GPC showed a moderate increase (~2.47%) in average resonant frequency values when compared to 7 days values as specimens aged with developing microstructure between seven and 28 days. Compared to this, PCC showed a smaller increase (~0.74%) possibly due to almost fully completed hydration within seven days.

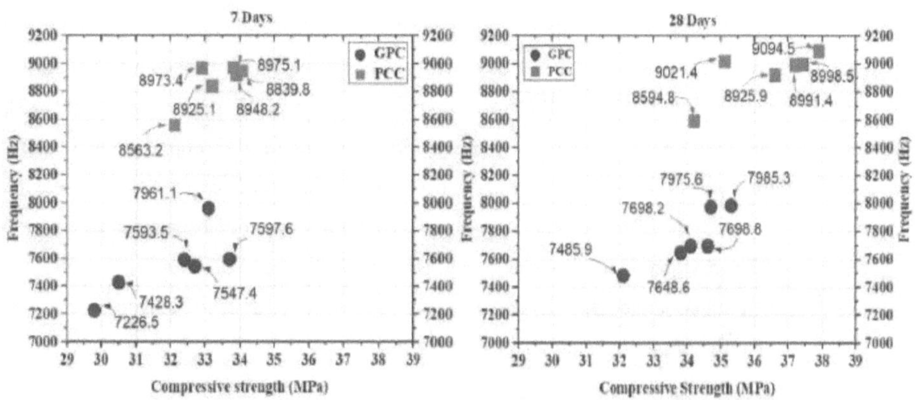


**Figure 5. Compressive strength of GPC and PCC vs. Frequency (7 & 28 days)**

Typically, the resonant frequency increases as the compressive strength increases. That is why, due to the higher compressive strength, PCC samples showed higher resonant frequency ,as shown in Figure 5, compared to its counterpart GPC at both 7 and 28 days.

It also can be seen in Figure 5 that the resonant frequency of GPC ranges from 7000 to 8000 Hz, whereas for PCC ranges from 8000 to 9100 Hz. The calculated standard deviation of GPC and PCC was 220.65 and 144.86, respectively, at seven days, whereas the calculated standard deviation of GPC and PCC was 178.69 and 161.18, respectively, at 28 days. This phenomenon is in good-agreement with previous studies performed at the University of Victoria [84,139] by the authors, where with average compressive strength of 20–30 MPa and the average longitudinal resonant frequency of PCC and GPC was about 9000–9500 Hz and 5500–6000 Hz, respectively. According to a study by Massoud et al. [140], GPC specimens have a lower frequency than PCC for mixes that have a similar average 28 days compressive strength due to elevated curing condition and internal moisture loss.

Equation (2) was used to calculate the elastic modulus of GPC and PCC samples at an age of seven days and 28 days. Then, a correlation between compressive strength and modulus of elasticity was derived (Figure 6). It can be seen in Figure 6 that a slight overall increase in compressive strength of PCC resulted in a much larger increase in the overall value of dynamic modulus of elasticity approximately ranging from 27 GPa to 33 GPa. However, GPC showed a minor increase of the dynamic modulus of elasticity, approximately ranging from 20 GPa to 24 GPa, with an increasing of compressive strength from 30 MPa to 35 MPa.

The standard deviation value of GPC and PCC was 1.24 and 0.76, respectively, at seven days, while the standard deviation value of GPC and PCC was 0.85 and 0.96, respectively, at 28 days. This is a very interesting finding and is in good agreement with Fernandez-Jimenez et al.'s study [141] where with a similar compressive strength the elastic modulus of fly-ash based GPC was in the range of 10–20 GPa, while cement-based concrete had a higher elastic modulus in the range of 25–35 GPa. According to Duxson et al. [142] and Sofi et al. [143], the elastic modulus of GPC depended on the $SiO_2/Al_2O_3$ ratio, mix proportion and curing method. However, both Sofi et al. [143] and Duxson et al. [142] reported that the elastic modulus of GPC varied from 10–55 GPa, which covers the values of elastic modulus of GPC reported in this research. The aforementioned issue indicates that Equation (2) is applicable for cement-based concrete, and this equation overestimated the dynamic elastic modulus of GPC owing to the higher resonant of PCC [84].

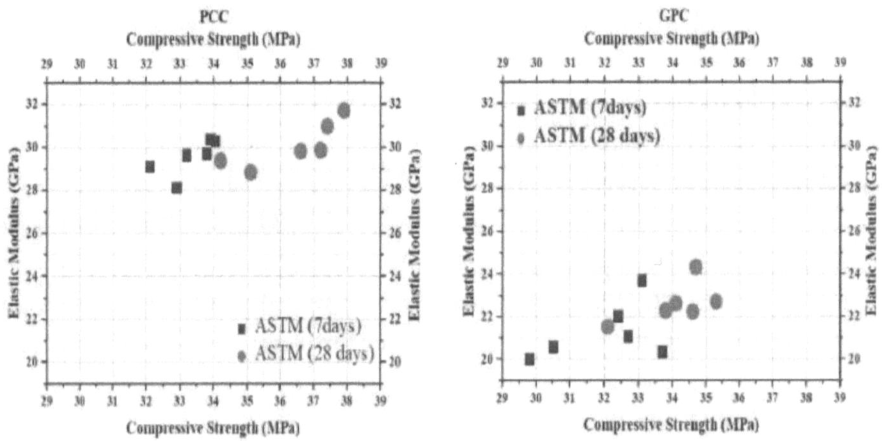

Figure 6. Elastic modulus versus compressive strength

It should be noted that the use of a higher amount of coarse aggregates in the mix proportion may increase/decrease the elastic modulus of concrete. In this study, due to the use of different types of concrete (GPC and PCC) and variation of the mix proportions, slightly higher content of coarse aggregates (4.36%) was used in GPC compared to PCC to achieve a compressive strength of 35 MPa. However, as is the case in this study, the elastic modulus of GPC is generally reported less than its counterpart PCC [141-143]. This phenomenon can be attributed to various factors including microstructure of concretes (porosity), strength, paste, elastic modulus of raw materials, characteristics of the transition zone, aggregates, etc. [79].

According to the Nath et al. study [55], the elastic modulus of ambient-cured GPC is about 25%–30% less compared to cement-based concrete at 28 days. It can be noted that both types of curing methods (heat curing and ambient curing) create GPC of lower elastic modulus compared to that of cement-based concrete.

In this study, material-focused investigation of both types of concrete showed that mixtures' stiffness/compressive strength increases with time (Figure 6). This is attributed to the increasing strength over curing time. However, in a concrete building after a brief period of time, this tendency becomes negligible due to the sustained loads (self-weight, dead loads and live loads) [144]. This is why it is significant to investigate the elastic modulus of materials when real loads are applied because the elastic modulus of concrete is relatively constant at low stress levels but starts decreasing at higher stress levels.

### 4.9.4 Comparison between predicted and experimental modulus of elasticity

In this study, the experimental results were compared with the predictive models (Table 6) proposed by other researchers to identify the best suited model for this type of mixture containing bottom-ash.

Table 6. Equations and establishers

| Equation | Establisher |
|---|---|
| $E_c = 2707 \sqrt{f_{cm}} + 5300$ | Hardjito et al. |
| $E_c = 3320 \sqrt{f_{cm}} + 6900$ | ACI 363 (Hardjito and Rangan) |
| $E_c = 3421 \sqrt{f_c}$ | Tempest et al. |
| $E = 840 - 886\sqrt{f_c} + 647f_c$ | Prachasaree et al. |
| $E = 2900f_c^{3/5}$ | Thomas and Peethamparan |
| $E = 4400\sqrt{f_c}$ | Thomas and Peethamparan |
| $E_c = f_c^{1.6412} \, 49.968$ | Wardhonoet al. |
| $E_c = \rho^{1.5} \, 0.024\sqrt{f_c}$ | Gunasekara et al. |

The experimental results for average modulus of elasticity (from experimental results) of six GPC specimens at an age of 7 and 28 days are indicated in Figure 7. It should be noted that all of the proposed models considered in this study are for GPC made only with fly-ash. However, bottom-ash has similar physical, chemical and mechanical properties as fly-ash [145]. One would expect that these models are also applicable for GPC made using a combination of bottom-ash and fly-ash as well. Figure 7 shows the elastic modulus of GPC achieved from experimental results and predictive models.

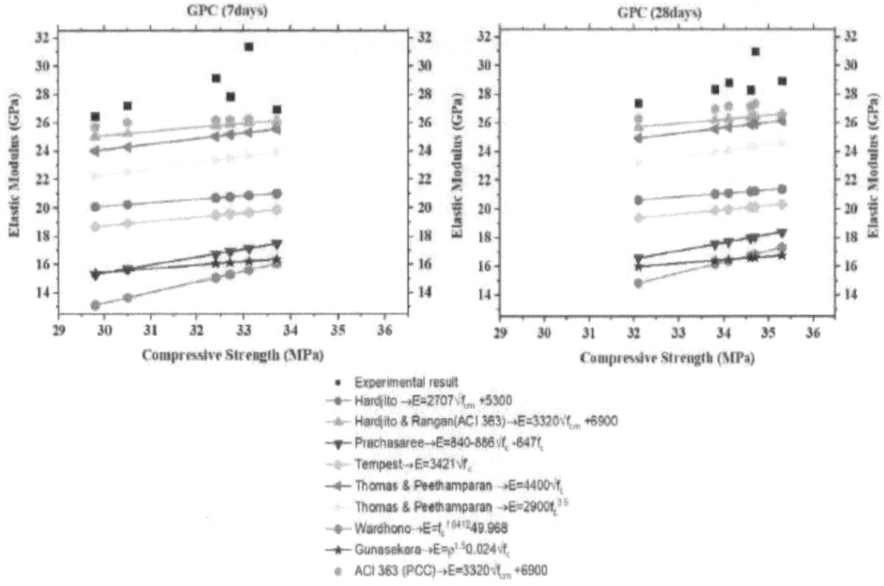



**Figure 7. Comparison between predicted and experimental modulus of elasticity**

The aforementioned Equations (3)–(10) were proposed to calculate the static elastic modulus from the compressive strength of GPC. It should be mentioned that the authors could not find any predictive models for the dynamic elastic modulus of K-based GPC made using a combination of fly-ash and bottom-ash. However, Yang et al. [36] and Wongpa et al. [146] reported that static predictive models can reasonably predict the dynamic modulus of elasticity because the predicted values are consistently lower or higher than the experimental values. Moreover, Yang et al. [36] used Kar et al.'s work [147] to develop a relationship between compressive strength and dynamic elastic modulus. According to their results, the maximum compressive strength difference between Yang's model [36] and Kar et al.'s model [147] was about 3.49 MPa when dynamic elastic modulus was about 23.69 GPa. Moreover, the values of static predictive models and the dynamic predictive models were in close agreement and had a similar trend.

In this study, an average of three readings were taken for each GPC sample to calculate the elastic modulus. A comparison between predictive models reported in previous studies for fly-ash-based GPC and the current study indicates that all models have a similar trend toward increasing elastic modulus with increased compressive strength. Even though there are some differences in compositions of GPCs, as can be seen in Figure 7, all of the predictive models achieved lower elastic modulus values than the experimental values in this study. The elastic modulus values obtained by Wardhono et al. [114] and Gunasekara et al. [117] are lower than experimental results of GPC made using fly-ash and bottom-ash. Authors attribute the low values in Gunasekara's model [117] to the effect of specific gravity of bottom-ash and fly-ash on density of GPC, which is lower compared to cement-based concrete. Authors believe that Wardhono's model [114] may not be very accurate due to the small sample size of data. As can be seen in Figure 7, the ACI 363 [108] model is the closest to the experimental results at both seven and 28 days. In the current study, the average calculated modulus of elasticity of GPC obtained by ACI's model is 8.88% (7 days) and 8.76% (28 days) lower compared to the respective average experimental results at the same age. Hence, the modulus of elasticity for GPC made using fly-ash and bottom-ash can equitably be calculated with reasonable correlation using Equation (4). Moreover, Figure 7 also shows that the model proposed by Thomas and Peethamparan [113] has an average elastic modulus of 11.68% (7 days) and 10.83% (28 days) lower than average experimental results at the same age. Hence, this model could also be used for this type of GPC with reasonable accuracy. It is well-known that ACI 363 [108] is applicable to PCC. However, the predictive model proposed by ACI 363 [108] was also used to make a comparison between the trend of GPC and PCC. It can be seen that the PCC values are slightly higher than GPC values. It shows that ACI 363 [108] marginally overestimated the elastic modulus of GPC. The authors propose the following empirical model to correlate the elastic modulus and compressive strength with a coefficient of determination ($R^2$) of 0.79.

$$E = 1.76 \, (f_c)^{0.64} \tag{11}$$

where $f_c$ has the unit of MPa, and the unit of "E" is GPa.

### 4.10 Conclusions

In this study, the effect of curing temperature on compressive strength and dynamic modulus of elasticity of K-based GPC was investigated. A combination of fly-ash and bottom-ash was used in order to develop a sustainable concrete mix. The comparative results of the study are summarized below:

1. Compressive strength increased approximately 3.5 times when steam-cured temperature increased from ambient to 80 °C. Compressive strength increased approximately 2.3 times when specimens were dry-cured at the same temperature range. Hence, steam curing is preferable to dry curing if rapid strength development is required.

2.   The measured values of the resonant frequency of K-based GPC (~7200–9000 Hz) with a compressive strength of 35 MPa was lower than that of cement-based concrete (~8000–9100 Hz) of 35 MPa due to the elevated temperature condition and internal moisture loss.

3.   The dynamic modulus of elasticity of bottom-ash based GPC was about 20–24 GPa at an age of 7 days and 28 days. Contrary to this, PCC showed a higher value of elastic modulus that ranged from 27 to 33 GPa.

4.   Several predictive models for fly-ash based GPC proposed by other researchers were used to find a model suitable for bottom-ash based GPC. The models proposed by ACI 363 [108] and Thomas et al. [113] were in good agreement with the average elastic modulus of GPC specimens in this study.

# Chapter 5   Microscopic Investigation of GPC

SEM is a test method that scans a specimen with an electron beam to create magnified images for investigation. The method is used very effectually in microanalysis and failure analysis of substances. SEM is performed at high magnifications, generates high-resolution images and precisely measures very small features and properties. In the present study, the authors wondered whether a new SEM method can be used to take an image and estimate the optimal parameters/properties of a frozen GPC sample. That is why, the GPC specimens were kept frozen in the SEM chamber to detect the effect of cryo on pores of GPC.

## 5.1   SEM images of GPC

Figure 8 shows the SEM of K-based GPC taken by a Hitachi S-4800 apparatus with 15 kV accelerating voltage. The production of GPC including mix design and curing method has been already explained in section 4.6.1. and 4.6.2. After 28 days of curing, the GPC specimens were broken into small pieces for further study using SEM. The specimens with an approximate dimension of 1.5×1.5 cm were randomly selected and placed into a vacuum desiccator to dry. After that, the surface of three specimens was coated with carbon to dissipate excess charge from the specimens. As can be seen in Figure 8(a), the microstructure of the K-based GPC is heterogeneous and porous with a small amount of unreacted by-products particles. Figure 8(b) shows the micro-cracks on the surface of K-based GPC which were probably initiated by the steam curing leading to shrinkage of the paste. This happens because capillary tension is generated over the depletion of water from paste during the polycondensation process [148].

**Figure 8. SEM images of GPC**

**(a) Unreacted particles at after 80 °C of curing (b) Micro-cracks**

## 5.2    Energy Dispersive X-ray Spectroscopy analysis

The Energy Dispersive X-ray Spectroscopy (EDX) detector was used to determine the most available elements inside the K-based GPC matrix. As can be seen in Figure 9, SEM images and EDX elemental mapping of GPC showed that distinctive elements of fly-ash and bottom-ash (Silicon (Si) and Aluminium (Al)) and K-based elements (Potassium (K)) in addition to Iron (Fe), Calcium (Ca) and Oxygen (O) mostly exist in the GPC paste (Figure 9). This result correlates with the EDS of the polished sample [149,150] meaning that EDX analysis of both polished and unpolished sample contains the phase of K-Si-Al.

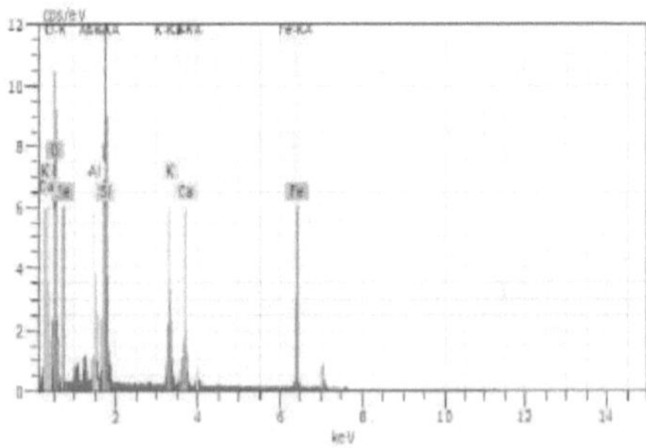

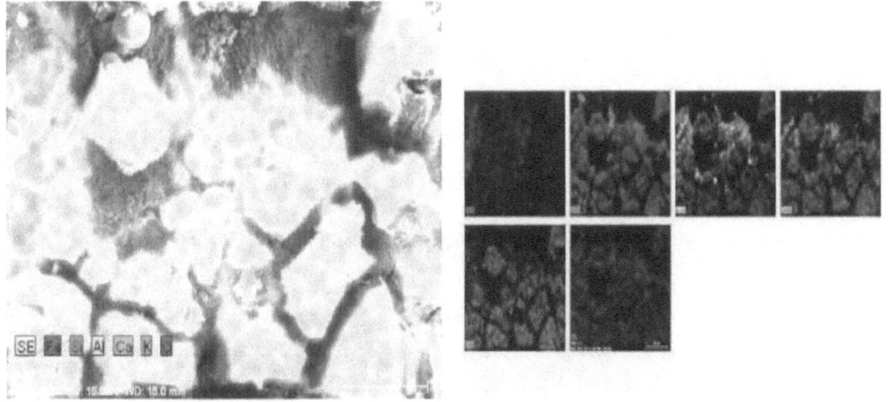

Figure 9. EDS analysis of GPC and raster images of GPC

## 5.3 Observational study on the effect of real environmental conditions on the morphology of GPC

In this study, two areas at the entrance of parking lot#3 at the University of Victoria were selected to place a total of 150 GPC and 210 PCC paver blocks exposed to real environmental conditions. To characterize the effect of aggressive conditions on the microstructure of GPC, small pieces of a GPC paver block was extracted after 90 days of exposure to real environmental conditions, to investigate the effect of real environmental conditions on change in the microstructure of the paste. Figure 10 shows the changes in morphologies of K-based GPC. As can be observed from Figure 10(a), the spherical shape of ash particles can be seen at the first day of exposure to real environmental condition but Figure 10(b) shows that the spherical shape of particles was damaged possibly due to temperature variation (9 °C to 2 °C) throughout exposure.

**Figure 10. SEM image of GPC**

**(a) First day of exposure to real environmental conditions**

**(b) 90 days of exposure to real environmental conditions**

## 5.4 Microstructure of GPC in a frozen condition

Th following manuscript has been published in the journal of "**Micron**". It should be noted that this chapter is an edited version of the published manuscript.

### 5.4.1 Introduction

GPC as a green material in the construction industry has brought a widespread of attention due to its ability to reduce $CO_2$ emission. GPC is an inorganic material where cement is replaced with industrial by-product materials and is activated by alkaline solution [20]. Freeze-thaw cycles in cold climate regions lead to cracks and spalls in the structure of GPC due to progressive expansion of its matrix. Deterioration increases as freeze-thaw cycles are reiterated, and the concrete structure gradually loses its durability [79]. Therefore, one of the challenges that can be highlighted in GPC related studies is freeze-thaw resistance of structure in sub-zero temperatures. Microscopic study of concrete plays a vital role in identification of nature of the binding between paste and aggregates. Consequently, this identification helps to researchers to increase performance of concrete under various conditions [151]. Since, there is a few literatures reported on microstructure of GPC [69,152], this aspect of K-based GPC is investigated in this paper.

Powers et al. [153] microscopically examined the effect of water on the interior walls of PCC and proposed the hydrostatic pressure theory of freeze-thaw failure deterioration. According to this theory, the conversion of water to ice creates 9% volumetric expansion and concrete fails if the hydrostatic pressure exceeds the tensile strength of the paste. Powers et al. [154] also proposed the osmotic pressure that the transformation of water to ice inside the large pores is faster than small pores. This solution concentration difference between large frozen pores and small unfrozen pores create osmotic pressure. So, the osmotic pressure damages the gel pores of the paste.

Hydration process creates lots of capillary and gel pores inside the microstructure of PCC. So, there is huge amounts of free spaces for water to penetrate through the saturated paste. However, water inside the capillary pores freezes below -12 °C. While, the temperature of -78 °C is needed to freeze water inside the gel pores. So, it can be concluded that frozen water inside the capillary pores cause the freeze-thaw failure damages of concrete [155]. Nevertheless, there is no literature reported on the domain of the freezing point in pores of GPC.

AEA is used for concrete mixtures exposed to the cold environments. AEA produces millions of tiny chambers for water to relieve internal pressure when the concrete experiences freeze-thaw cycles [156]. Sufficient amount of AEA is required for concrete matrix to resist frost actions. AEA improves workability of the concrete paste although ultimate strength of concrete paste decreases [79].

More air voids in the microstructure of concrete mixture provide more spaces to relieve hydrostatic pressure and prevent deterioration to concrete mixture [157,158]. Currently, there are abundant theoretical and experimental studies used to determine freeze-thaw resistance of PCC [159,160]. However, no experimental work is reported on characterizing cryo formation in microstructure of GPC due to unavailability of 4D-LTSEM equipment.

Sun et al. [161] investigated the effect of AEA on freeze-thaw resistance of fly-ash mortar and PCC in accordance with ASTM C666 [9]. The mass loss and dynamic elastic modulus retention of four compositions (PCC without AEA, PCC with AEA, fly-ash mortar without AEA and fly-ash mortar with AEA) were measured after 300 freeze-thaw cycles. Results showed that PCC without AEA subjected to the greater mass loss, while fly-ash with and without AEA gained mass after 300 cycles. AEA-based fly-ash mortar indicated a dynamic modulus loss of 6.8%, while non-AEA-based fly-ash mortar lost 8.4% of the elastic modulus after 300 cycles of freeze-thaw. The results also showed that non-AEA-based PCC resulting in a strength loss of about 20% and AEA-based PCC had a final loss of only 5% after 300 cycles.

Since, there is no research reported on the in-situ investigation of cryo formation of GPC, this research is targeted to experimentally and deeply investigate the morphology of cryo formation inside the entrained air voids of GPC matrix to have a comprehensive understanding of the extent of the microstructure of GPC paste exposed to the freeze-thaw cycles. Traditional microscopic devices are not able to maintain the temperature for a long enough period of time. So, tiny samples are expected to be warmed very quickly in the testing chamber. In this regard, 4D-LTSEM was used to take high-resolution images from ice propagation inside air voids of the GPC paste. 4D-LTSEM is setup to image tiny specimens kept under frozen conditions. This equipment has a temperature controlled stage so that the specimens can be held from -180 °C to +100 °C. Consequently, there is an opportunity to observe cryo products in air voids at a freshly fractured cross section.

### 5.4.1.1 Research contribution

Freeze-thaw resistance is a very critical characteristic of concrete when it is exposed to an aggressive environment. GPC structures located in cold environments can get damaged by freeze-thaw cycles during their life span. Hence, it is very important to study the microstructure of the paste in the frozen condition giving an insight into the cryo morphology in the air-voids of GPC. This is important as the morphology can dictate the amount of internal stresses caused in the paste and indicate the temperature range when substantial amount of cryo products start to form in the microstructure of the mixture. There is limited work that focuses on investigation of cryo morphology in entrained air voids of traditional concrete [162-164] but no research has been reported on formation of cryo products in air voids of GPC. This paper presents novelty focused on study of mechanism of cryo creation in the air-voids of GPC by using 4D-LTSEM.

## 5.4.2 Experimental work

### 5.4.2.1 Fly-ash and bottom-ash

The properties of fly-ash and bottom-ash have been already explained in section 4.5.1.

### 5.4.2.2 Morphology of fly-ash and bottom-ash

The morphology analysis of the two industrial by-products are performed by using SEM to obtain more details about the microstructure of raw fly-ash and bottom-ash. Both materials were analysed under 16.0mm×1.20K magnification.

Figure 11 shows that fly-ash consists of spherical particles with a regular smooth texture, showing that fly-ash particles were subjected to the high temperatures during coal combustion in furnace [165]. As it is reported, the particle size of fly-ash is about 1 to 20 µm with average size of 10 µm [20].

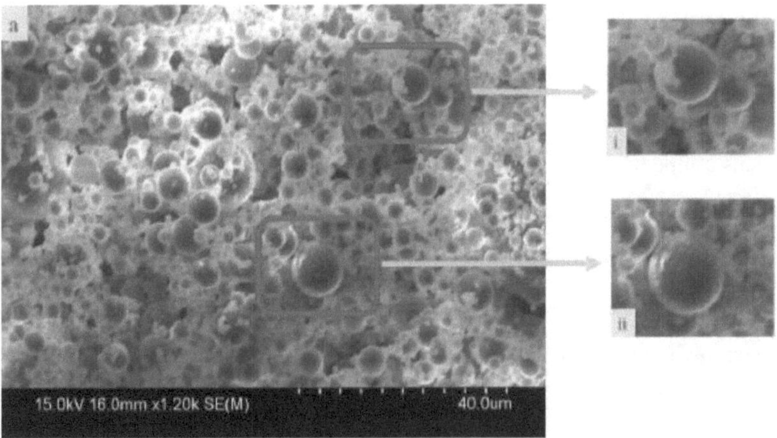


Figure 11. SEM image of fly-ash particle
(i)&(ii) Random areas

The SEM image of bottom-ash (Figure 11) indicates both regular and irregular shaped particles. Two particles shown in the red rectangle were randomly selected to measure the size of regular and irregular shapes grains. Mal'chik et al. [166] investigated the chemical content and particle size distribution of bottom-ash particles of 130 samples. The results show that the particle dimensions of bottom-ash vary from 10 to 100 μm. A comparison between Figure 11 and Figure 12 shows that bottom-ash particles are coarser than fly-ash particles. These coarser particles cause lower reactivity when used in GPC.

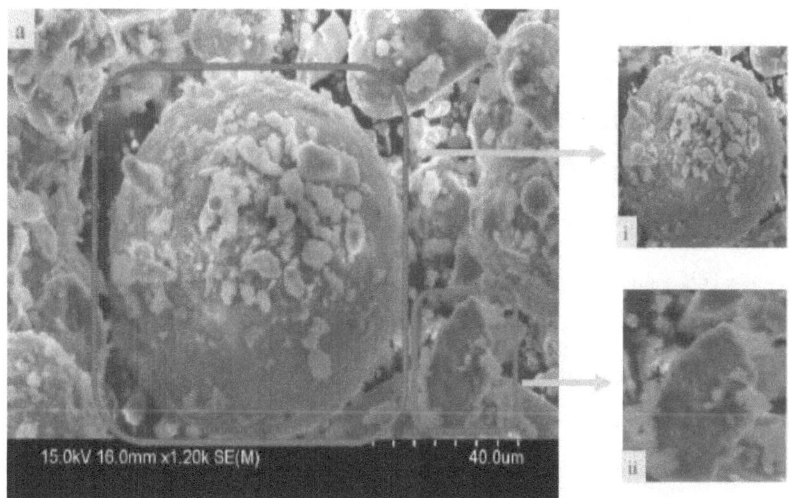


**Figure 12. SEM image of bottom-ash particles**
(i)      **Regular-shape**
(ii)     **Irregular-shape**

### 5.4.2.3 Alkaline solutions and aggregates

The properties of alkaline solutions and aggregates have been already explained in section 4.5.2.

In the frozen condition, the deterioration of GPC happens due to moisture freezing in air voids of mixture and subsequent expansion of the paste. The expansion force from the frozen water can exceed the tensile strength of the GPC and delaminate the paste. This phenomenon destructively affects the ultimate strength and durability of GPC and threatens the lifespan of a structure [79]. Generally, the resistance of GPC subjected to the freeze-thaw cycles is dependent on content, size and spacing of air content [167]. AEA as one of the surface active agent/ surfactant is basically used to relieve internal pressure and increase the freeze-thaw resistance of GPC. AEA entrained air bubbles in mixture by changing the surface tension of water and act at the air-water interface within the mixture. AEA consists of a hydrophilic head (polar) and a hydrophobic (radical) tail, by which it gets orientated into the air phase with the polar molecules inward toward the water [79]. AEA creates more tiny spaces for the expansion of ice by producing air bubbles with diameter ranging from 10 μm to 1mm [167].

In this study, the air content of GPC was measured by pressure apparatus with a capacity of 0.25 ft$^3$ in accordance with ASTM C 231 [168]. The fresh GPC was poured in a pressure vessel in three layers. Each layer was consolidated by a 25 tamping rod-drop from a height of 30 mm. After dropping, the measurement in a pressure apparatus was conducted. The average air content of six samples of fresh AEA-based GPC was about 4.5±0.5%.

To produce GPC samples, dry materials were first mixed in a mixer for 1 minute. Then, KOH and K$_2$SiO$_3$ solution, already prepared (24 hours before casting day), were added steadily to the mixture along with extra water for 3 minutes, followed by a 3 minutes rest period, followed by 2 minutes of final mixing. After molding, GPC samples were vibrated for 30 seconds to discharge the air bubbles to the surface. The samples were then kept at an ambient temperature for 24 hours (approximate relative humidity range of 45% to 70% and approximate temperature range of 5°C to 15°C). The samples were demolded after 24 hours and were steam cured at 80 °C for 24 hours [83].

A tiny chunk of the paste from the surface of 28-day cured GPC samples were taken to study the pore walls in the 4D-LTSEM cooling chamber. Liquid nitrogen was used to freeze the GPC specimens. 4D-LTSEM is designed to take images at very low temperature of upto -180 °C. In this study, initially, the samples were frozen at a temperature of -180 °C to take the images of cryo products and then the samples were slowly warmed in the chamber to take images from the exact same locations after the sublimation of ice.

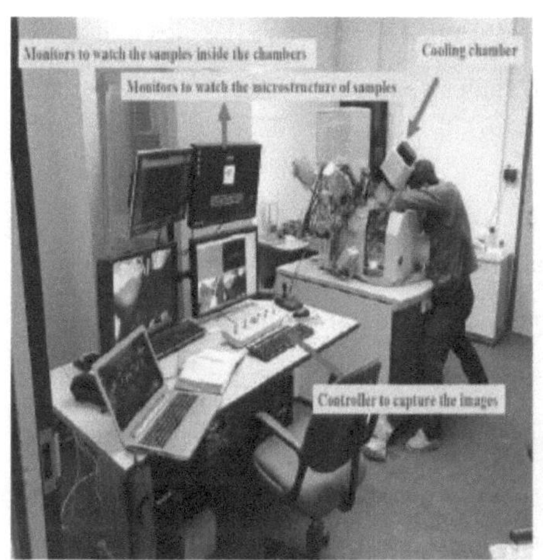

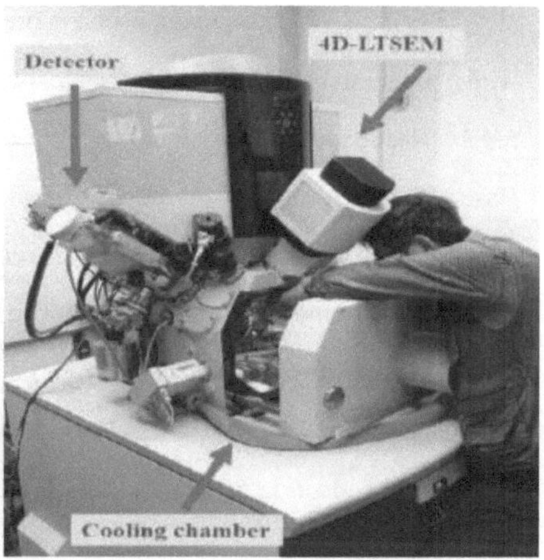


**Figure 13. 4-D Monitoring section (left image), 4-D LTSEM (right image)**

The 4-D LTSEM (FEI Helios Nano-Lab Dual-Beam 650) is a combined field emission SEM and Focused Ion Beam (FIB) technology. It is designed to capture high resolution SEM images at low temperatures (up to -180 °C) and high temperatures (up to +100 °C). The equipment has low kV operation down to 500 V and up to 65 nA beam current. The main part of 4-D LTSEM shown in Figure 13 is the cooling chamber. This chamber/stage is able to maintain the temperature for the specimen that needs to be tested at sub-zero temperature.

Isac-García et al. [169] proposed that the rate of sublimation decrease when the pressure in the chamber becomes higher than the pressure of the ice. Sublimation continuously begins from external surface and migrates to inside the specimens. This phenomenon creates two different layers in a specimen, dry layer and frozen layer. The boundary of these two layer named "sublimation interface" or "ice interface" [169]. So, as it was suggested by Corr et al. [163], in this study attempts have been made to keep the average chamber pressure at about $1.3^{-9}$ millibar to reduce the rate of sublimation of water from surface of specimen and to avoid creation of sublimation interface to detect the main ice morphology inside the air voids of GPC.

Additionally, in real environmental conditions, since the gas molecules sublimated from dry layer need to be fitted in the atmospheric gases, the process of sublimation is preferred to be slow in barometric pressure/atmospheric pressure [169]. ASTM C666 [9] is also well-agreed with a longer transition period from freezing to thawing phase.

### 5.4.3 Results and discussion

Basically, the formation of the GPC paste is totally different from that in PCC which consequently leads to a different microstructure [69,151]. Figure 14 is a micrograph of K-based GPC i.e. the SEM image at 600X magnification. The SEM image of K-based GPC shows a large proportion of reacted particles. So, relatively dense matrix can be seen in microstructure of GPC. It also can be observed that both bottom-ash and fly-ash particles are properly embedded into the paste. So, it means that GPC pastes were fully and uniformly cured.

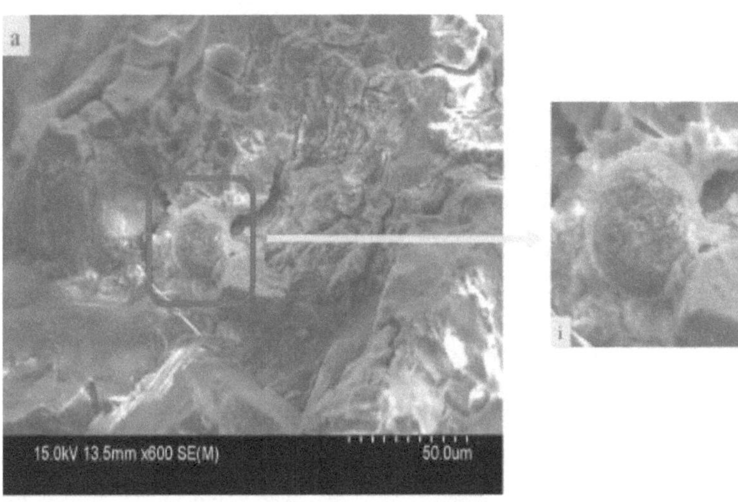


**Figure 14. SEM image of GPC**
**(i)        Unreacted particle of by-products**

From visual observation under the 4D-LTSEM, it was found that the air void structure of GPC is totally different from air voids reported for PCC [163,164]. Initially, the samples were frozen at the temperature of -180 °C to take images of created cryo products and then the samples were warmed to take image after sublimation of cryo products. As can be seen in Figure 15(a), the surface of paste is totally covered with cryo products at temperature of -180 °C and most of the air voids are invisible. So, all the voids are probably clogged with cryo products and possibly the paste is subjected to significant expansion. Figure 15(b) indicates the same area after sublimation of the cryo products at a temperature of about -70 °C. At this temperature, due to the high freeze-resistance capacity of the air voids of GPC, most of cryo products disappeared and more voids are visible on the surface of the paste. The images taken at a temperature of -18 °C -10 °C (Figure 15(c) and (d)) indicates that the surface of the paste is almost smooth, and the voids are completely empty. The red rectangles in Figure 15(c) and (d) were randomly selected to observe the rate of cryo products sublimation. As it can be seen, there is no evidence of cryo products inside the random area due to high freeze-resistance of the paste.

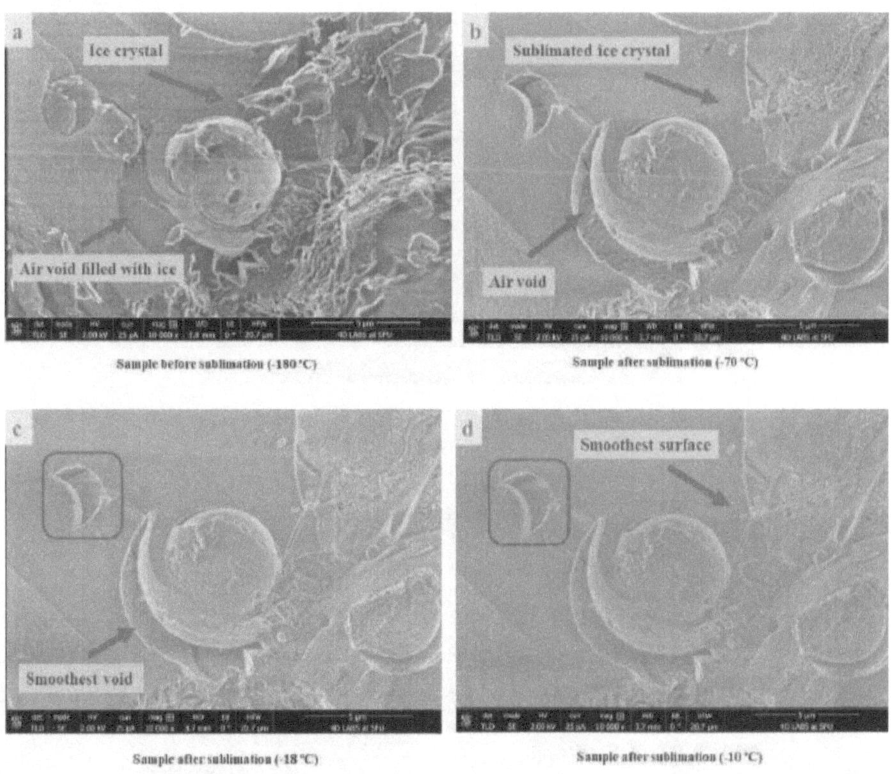


**Figure 15. 4D-LTSEM images of frozen GPC sample at following temperatures: (a) -180 °C, (b) -70 °C, (c) -18 °C and (d) -10 °C**

Generally, unlike the cryo formation in the air voids of PCC reported by David J. Corr [163], the ice on the surface of air voids formed non-discretely; ice covered the air voids because the samples were immersed in the water to be fully-saturated. Based on the ASTM C666 [9], at the end of the cooling period (-18 °C) the paste is still expected to show proper strength. The comparison between the four micrographs in Figure 15 shows that the rate of cryo formation from a temperature of 0 °C to -10 °C is same as cryo formation from temperature of -10 °C to -18 °C. Hence, it can be concluded that only water inside the capillary pores froze below -18 °C. However, from visual observation, it was found that the rate of cryo products sublimation decreased dramatically from -70 °C to -18 °C. It shows that the ice inside the gel pores began to melt at a temperature of -70 °C to -18 °C.

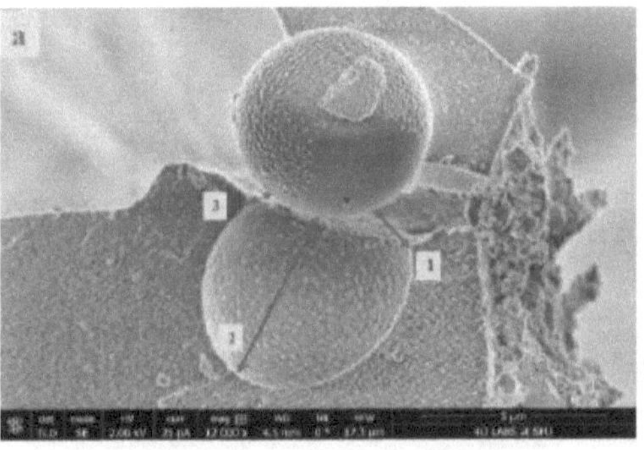

Sample before sublimation (-180 °C)

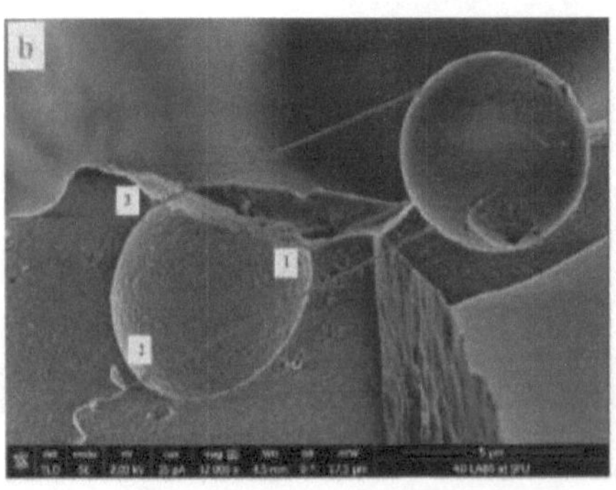

Sample after sublimation (-70 °C)

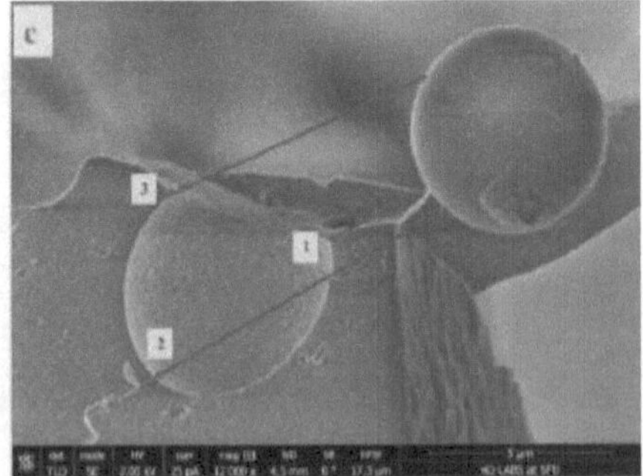

Sample after sublimation (-10 °C)

**Figure 16. 4D-LTSEM images of by-product particle at following temperatures: (a) -180 °C, (b) -70 °C and (c) -10°C**

Figure 16 also shows sublimation process of ice crystals of GPC from temperature of -180 °C to -10 °C. It is obvious in Figure 16(a) that ice totally covered the paste at a temperature of -180 °C and ice was dramatically sublimated from surface of paste by decreasing temperature from -180 °C to -70 °C as shown in Figure 16(b). Eventually, most amount of ice sublimated from the paste at a temperature of -10 °C as shown in Figure 16(c). However, Figure 16 also shows that the rate of ice sublimation between -70 °C to -10 °C is higher compared to rate of cryo products sublimation between -180 °C to -70 °C. So, it shows that GPC is frost resistant up to -70 °C.

D.J. Corr [163] proposed that contact angles between cryo products and air void wall is a significant parameter to be measured because it affects the hydraulic pressure of the capillary pores during the freezing of water. However, in this study, it is found that cooling temperature also causes by-products particle to move apart from paste. An image (magnification of 12K) from one particle was taken to measure the distance of the particle from three reference points (1, 2, 3). Table 7 indicates that the average distance of the particle at -180 °C is 2.18 μm. However, when the temperature increased from -180 °C to -70 °C, the average distance of the particle increased from 2.18 μm to 6.68 μm. Table 7 also shows that there is no considerable change in distance when temperature increased from -70 °C to -10 °C.

Table 7. Distance measurements

| Figure 16 code | NO. | Distance (μm) | Average (μm) |
|---|---|---|---|
| a | 1 | 1.29 | |
| | 2 | 4.89 | 2.18 |
| | 3 | 0.38 | |
| b | 1 | 3.24 | |
| | 2 | 9.19 | 6.68 |
| | 3 | 7.63 | |
| c | 1 | 3.25 | |
| | 2 | 9.21 | 6.69 |
| | 3 | 7.63 | |

Generally, it is reported that elevated curing temperature is helpful toward dissolution of by-products particle [77,170]. GPC requires heat treatment in order to obtain high compressive strength. It means higher curing temperature results in higher interfacial transition zone bonding. In contrast, Pilehvar et al. [11] proposed that frost actions cause microcracks initiation and bonding strength loss at the weak interfacial transition zone.

Figure 16 and measured distance (Table 7) show that by-product particle located on an air void at temperature of -180 °C. However, after sublimation to -70 °C, the particle moved away from its location. This phenomenon also can be explained with bonding strength of paste at lowest temperature. At temperature between 0 °C to -70 °C, most likely bonding strength of GPC was higher than that at temperature of -180 °C and only freezing of water in larger pores was predominant. However, the bonding strength of the paste decreased at a temperature between -70 °C to -180 °C and probably freezing of water in smaller pores was predominant due to the drop of freezing point of the capillary water. In this study, the distance between the particle and matrix decreased with the increasing temperature from -180 °C to -10 °C. Generally, at large distances, the strength reduces, meaning that there is no interaction between particles. At short distances between the two particles, due to attraction force, and chemical tendency of particles to create a polymer/covalent bond, the bond strength between particles is very high.

### 5.4.4 Conclusion

In this paper, K-based GPC as a sustainable material was developed to investigate the microstructure of cryo products structure in the air voids of paste. A new technique was also used to maintain the temperature as low as possible to capture morphology of cryo products propagation in the air void of GPC. The results indicate that:

1- The SEM image of K-based GPC shows that due to full and uniform curing process, most proportation of fly-ash and bottom-ash reacted with alkaline solution and created a dense matrix.

2- Investigation of cryo formation in the air voids of GPC showed that the cryo non-discretely covered the surface of air void because the specimen were fully-saturated.

This is not in full agreement with literature where cryo formation in the air voids of PCC indicated discrete and hemispheric cryo crystals along the wall of air voids.

3- Microstructure investigation of cryo products showed that cryo products covered GPC at temperature of -180 °C but most amount of cryo products was sublimated at a temperature of -70 °C. Finally, a few amount of cryo products was observed on surface of the paste at a temperature of -10 °C. However, it was found that the rate of cryo products sublimation between -180 °C to -70 °C was high. While, due to frost resistance of GPC upto -70 °C, there was no significant sublimation rate when temperature changed from -70 °C to -10 °C.

4- Distance measurement of GPC was performed to investigate the effect of cooling temperature on the bonding strength of the paste. The average distance measurement of particle at -180 °C, -70 °C, and -10 ° C showed that the average distance of the particle at -180 °C was 2.18 µm. However, by increasing temperature from -180 °C to -70 °C, the average distance of the particle increased from 2.18 µm to 6.68 µm. It was also observed that there was no significant change in distance when sublimation process occurred from -70 °C to -10 °C. So, it can be noticed that GPC has reasonable bonding strength up to -70 °C.

# Chapter 6    Freeze-Thaw Resistance and Leach-ability of GPC

Developing GPC as a construction material raises the question of sufficient durability and performance in cold regions. Moreover, limited studies have been done on freeze-thaw durability of K-based GPC made by 50% fly-ash and 50% bottom-ash. This was the motivation for this research that studied the freeze-thaw resistance of fly-ash based GPC and bottom-ash based GPC based on ASTM C666 [9]. Moreover, application of by-products in construction materials is limited by the potential risk of toxic metals leaching. There is also limited data available on leach-ability of K-based GPC. In this study, leach-ability of GPC was also measured to investigate its toxicity characterization.

## 6.1    Investigation of freeze-thaw durability and leach-ability of K-based GPC

The following manuscript has been published in the **"Journal of Composites Science"**. It should be noted that this chapter is an edited version of the published manuscript.

### 6.1.1    Introduction

Fly-ash and bottom-ash are by-products of the combustion of pulverized coal in thermal power plants. These by-products are pozzolans and typically consist of Silicon Dioxide ($SiO_2$), Aluminum Oxide ($Al_2O_3$) and Iron Oxide ($Fe_2O_3$) [171]. It is projected by Miller et al. [172] that coal production will increase up to 1000 million tons annually by 2040. So, due to the environmental regulations, utilization of by-products such as fly-ash and bottom-ash in different applications must be provided to protect natural resources and avoid landfill disposal of ashes. Although, broadly speaking, the use of by-products in various industries leads to a cleaner environment, the Environmental Protection Agency [173] reported that by-products can pollute groundwater and can increase a person's health risk to incurable diseases. So, besides creating a durable concrete made using by-products, it is worthwhile to assess the environmental impact of concrete.

PCC is one of the most used construction materials all over the world due to its high durability, high mechanical properties, and long service life. However, the production of cement as a main constituent of PCC requires energy that leads to the generation of $CO_2$. According to the Portland Cement Association (PCA), 1000 kg of Portland Cement Production releases 927 kg of $CO_2$ into the atmosphere [174]. Hence, to alleviate this impact on the atmosphere, it is critical to reduce the use of Portland cement in the concrete by replacing it with other pozzolanic materials such as fly-ash, slag and bottom-ash.

Due to the issues mentioned above, attempts have been made to use waste materials such as fly-ash to make durable concrete. However, the use of bottom-ash to produce concrete is not widely reported in the literature. French scientist Joseph Davidovits initially found one such concept, that of the "Geopolymer". Then, slag-based geopolymer cement was made in the 1980s [175]. Subsequently, in 1997, based on the obtained results for slag-based geopolymer cement, Silverstrim et al. [176], Van Jaarsveld et al. [5] created an inorganic polymer concrete called GPC. GPC is produced by reacting aluminate and silicate bearing materials with a caustic activator [20]. Since then, numerous works have reported on the development of Na-based GPC durability [7,8,75,82]. However, there are only limited studies conducted on the development of mix design, durability and service life of K-based GPC made by the combination of bottom-ash and fly-ash. Habert et al. [74] compared the environmental impact of GPC made by two types of by-products (fly-ash and slag), metakaolin and PCC using the LCA methodology. The results confirmed that GPC has a lower environmental impact in terms of emission of $CO_2$.

Freeze-thaw damage is a potentially critical deterioration mechanism that occurs not only in cement-based concrete but also in GPC structures. The reason for the deterioration of GPC in a cold condition is the water in cracks/capillary pores. The water in the cracks/capillary pores converts into ice when the material was exposed to the freeze-thaw cycles, where this issue leads to the internal expansion stress in the paste. This expansion stress causes internal micro-cracks. Continuously, an increase of freeze-thaw cycles leads to an extension of micro-cracks. Fu et al. [177] studied the freeze-thaw resistance of Na-based alkali-activated slag concrete. Six specimens of five mix proportions were cast and tested to measure the freeze-thaw resistance of alkali-activated slag concrete. Mass and Relative Dynamic Elastic Modulus of Elasticity (RDME) of each sample were measured after 25 freeze-thaw cycles. The result of the freeze-thaw test showed that RDME and mass of all mix proportions decrease about 10% and less than 1% after 300 cycles of freeze-thaw respectively. The authors also developed damage mechanics-based models using RDME. The authors reported that attenuation and power function models are more suited to accumulative and exponential damage models respectively.

This paper deals with the leach-ability of heavy metals of GPC. Basically, the potential leaching of heavy metals into the groundwater is a crucial concern with the use of by-products as a constituent of GPC, especially once the by-products are mixed with chemical materials such as KOH and $K_2SiO_3$. Thang et al. [178] investigated the leach-ability of hazardous metals including Copper, Cadmium, Lead, Iron and Chromium of fly-ash based and red-mud-based geopolymer. Nine mix proportions were made to use Inductively Coupled Plasma-Atomic Emission Spectrometry (ICP-AES) at a pH of 7 to characterize the heavy metals in the geopolymer. The authors reported that raw materials leached a high amount of Lead, Palladium, Chromium and other hazardous metals. However, at a pH of 7, the level of leached metals from geopolymer materials was within the range specified by the European Standard. It is also found that geopolymer materials had lower concentrations than their constituent raw materials (fly-ash and red-mud). Arioz et al. [49] studied the leach-ability of fly-ash based geopolymer paste cured at different temperatures, including 40 °C, 80 °C and 120 °C for 6,

15 and 24 hours. The hazardous metals including Arsenic, Lead, Chromium, Cadmium and Mercury were characterized using TCLP test for samples and ICP-AES for solutions extracted from GPC and fly-ash. The results of the TCLP test showed that fly-ash has higher heavy metals concentration than geopolymer paste. Moreover, the higher concentration of Arsenic and Mercury were measured in the solution of the samples cured at 120 °C for 15 and 24 hours.

#### 6.1.1.1  Research significance

First of all, ample studies on the durability and service life of Sodium-Based (Na-based) GPC have been reported. However, limited studies have been reported on the mechanical properties of Potassium-Based (K-based) GPC. Moreover, it is reported that bottom-ash has similar physical, chemical and mechanical properties to fly-ash [145]. Since the use of bottom-ash is relatively limited, dealing with this industrial material is now posing to be one of the most significant challenges in recent years. So, in the current study, in meso-scale level, attempts have been made to produce an appropriate mix proportion for K-based GPC made by a combination of 50% fly-ash and 50% bottom-ash (here noted as bottom-ash based GPC for further descriptions). K-based GPC synthesized only with fly-ash was also produced to compare its freeze-thaw resistance with bottom-ash based GPC.

Freeze-thaw delamination and heavy metals leaching are amongst the most persistent concerns of material in a frozen condition. Basically, aggressive environments damage the structure of GPC and consequently decrease their life span. Since there are limited studies that have reported on examination of the durability of K-based GPC, it is important to investigate the freeze-thaw resistance and leach-ability of K-based GPC synthesized by both fly-ash and bottom-ash.

Regarding freeze-thaw resistance properties, numerous predictive models were proposed for Na-based GPC [75,179,180]. Hongfa et al. [181] proposed a damage model for cement-based concrete. While, authors could not find any empirical model for K-based GPC, this model was used for K-based GPC in the current study. Moreover, the damage variable of K-based GPC was calculated and compared with other research studies to find the applicability of this freeze-thaw resistance model for by-products-based GPC [40,182].

### 6.1.2   Experimental work

#### 6.1.2.1   Precursors

The properties of precursors including fly-ash, bottom-ash and alkaline solutions have been already discussed in sections 4.5.1 and 4.5.2 respectively. A couple of more information about solutions is as follows:

Basically, a combination of alkali solution and soluble silicate is needed to produce GPC with desire durability. Since the durability of K-based GPC has only been reported by a few researchers [183,184], the combination of KOH and $K_2SiO_3$ was used in the present study because at elevated temperature (higher than 30 °C, GPC made by K-based is more steady than Na-based GPC in terms of mechanical properties including compressive strength [30]. KOH flakes obtained from Sigma-Aldrich Private Ltd., Missouri, United States, and $K_2SiO_3$ powder (AgSil 16) obtained from PQ Corporation (USA) were used in this study. Chemical elements of $K_2SiO_3$ obtained from the Material Safety Data Sheets (MSDS) of the product are shown in Table 8. The specific gravity of KOH and $K_2SiO_3$ was 1.45 and 1.26, respectively.

**Table 8. Chemical composition of $K_2SiO_3$**

| Compound | $K_2O$ | $SiO_2$ | $H_2O$ |
|---|---|---|---|
| %W/W | 32.4% | 52.8% | 14.8% |

#### 6.1.2.2   Microstructural study of fly-ash and bottom-ash

The SEM of fly-ash and bottom-ash was investigated using Hitachi S-4800, Chiyoda, Tokyo, Japan at the Advance Microscopic Facility (AMF) of the University of Victoria. The SEM operated at an accelerating voltage of 15 kV. Both fly-ash and bottom-ash were analyzed under 16.0 mm × 900 magnification.

The SEM images of both fly-ash and bottom-ash are presented in Figure 17(i,ii). Figure 17(i) shows that the fly-ash particles are spherical in shape and hence known as cenospheres (perfectly round smooth and intact) with the presence of a few irregular particles. Figure 17(ii) shows the bottom-ash particles that are larger in size and sub-angular to angular in shape. It also can be seen that bottom-ash particles are porous with tiny pores visible in Figure 17(ii). This porous nature of bottom-ash causes bottom-ash to absorb more water than fly-ash [71]. This needs to be properly accounted for when using bottom-ash in GPC production since excess water might have a negative impact on the GPC properties.

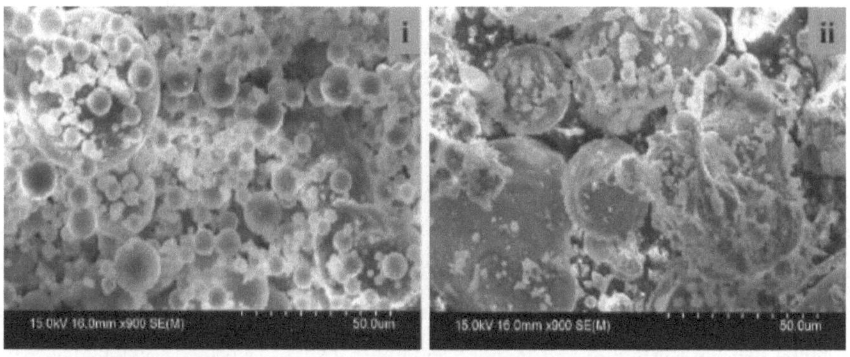

Figure 17. SEM images of fly-ash (i) and bottom-ash (ii)

### 6.1.2.3   Aggregates

The properties of aggregates have been already explained in section 4.5.2.

### 6.1.3   Mix design and specimen preparation of GPC

The mix proportion of GPC shown in Table 9 was derived after extensive initial trial experiments performed in the Facility for Innovative Materials and Infrastructure Monitoring (FIMIM) of the University of Victoria, Canada by the authors [36,82,83]. The production of GPC samples has been already explained in section 4.6.1.

Table 9. Mix design

| Material | Fly-ash based GPC (Kg/m$^3$) | Specific gravity |
|---|---|---|
| Fly-ash | 388 | 2.5 |
| Bottom-ash | 0 | 2.3 |
| Coarse aggregates | 1170 | 2.69 |
| Sand | 630 | 2.60 |
| KOH (12M) | 85.16 | 1.45 |
| K$_2$SiO$_3$ | 125.74 | 1.26 |
| Extra Water | 38.71 | 1 |
| AEA | 1.5 | 1.03 |
| Total | 2439.11 | 14.83 |

### 6.1.4   Curing of GPC samples

Several efforts have been made for characterizing the influence of curing environments on different properties of Na-based GPC [107,185,186]. The curing methods of GPC have been already explained in section 4.6.2. According to the results, steam-cured GPC samples showed greater compressive strength at temperature of 80 °C for 24 hours. So, the steam curing method and temperature of 80 °C were used to cure 54 cylindrical fly-ash based and bottom-ash based GPC samples (100 × 200 mm) for this study. More details about the curing regime used are available in prior studies [36,82,83].

### 6.1.5  Methodology

Figure 18 shows the scope of work of this study. In this study, first of all, attempts have been made to produce bottom-ash based GPC and fly-ash based GPC using the steam curing method (at a temperature of 80 °C). After that, cylindrical bottom-ash based GPC and fly-ash based GPC specimens were made to measure their compressive strength. In meso-scale level, the freeze-thaw resistance, resonant frequency and leach-ability of bottom-ash based and fly-ash based beams were then tested using the NDT. Eventually, a comparative study was made between concrete parameter, compressive strength and number of cycles. It should be noted that the leach-ability of bottom-ash based GPC was separately studied using the TCLP test.

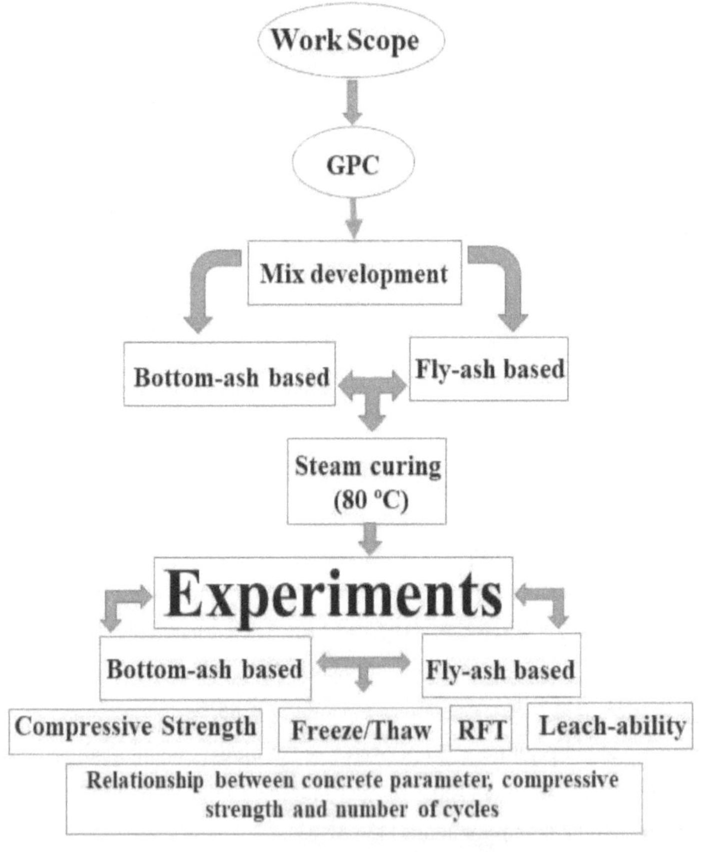


Figure 18. The work scope of the current study

### 6.1.5.1  Compressive strength

The steam-cured fly-ash based and bottom-ash based GPC samples (100 mm diameter and 200 mm height) were tested at the age of 28 days in accordance with ASTM C39 [105] using Forney compression testing machine model #AD 650.

### 6.1.5.2  Freeze-thaw test

In order to verify the durability of GPC under cold weather conditions, the freeze-thaw test was performed. Six GPC prisms (76 × 102 × 406 mm) were exposed to 300 freeze-thaw cycles, with temperature operating range from −17.8 °C to +4.4 °C and a humidity range from 10% to 95% in accordance with ASTM C666 [9]. Procedure "A" was selected in this study which arranges rapid freezing and thawing in water. After every 30 freeze-thaw cycles, samples were pulled off from the freeze-thaw cabinet to measure their mass loss, RDME and leaching.

### 6.1.5.3  Dynamic elastic modulus

One of the objectives of this research was to calculate the RDME using an NDT called RFT/Resonant Frequency Gauge (RTG). Figure 19 shows the components of the RFT/RTG device used for this study. Firstly, an accelerometer, with a frequency response measurement range of 20,000 Hz, was attached to the GPC surface using adhesive grease. After attaching the accelerometer and positioning GPC samples to the required mode of testing, a standard ball tip hammer weighing $110 \pm 2$ g with a tip diameter of 10 mm, is used to strike the surface at precise locations on the samples being tested. The achieved time domain signal was amplified and passed through BNC connection/cable. Finally, Olson instruments' RTG software records/shows the resonant frequency. ASTM C666 [9] suggested the following equation to calculate the RDME:

$$P_c = (\frac{n_1^2}{n^2}) \times 100 \tag{12}$$

where:

$P_c$ = RDME, %.

n = fundamental transverse frequency at 0 freeze-thaw cycles.

$n_1$ = fundamental transverse frequency after 'n' freeze-thaw cycles.

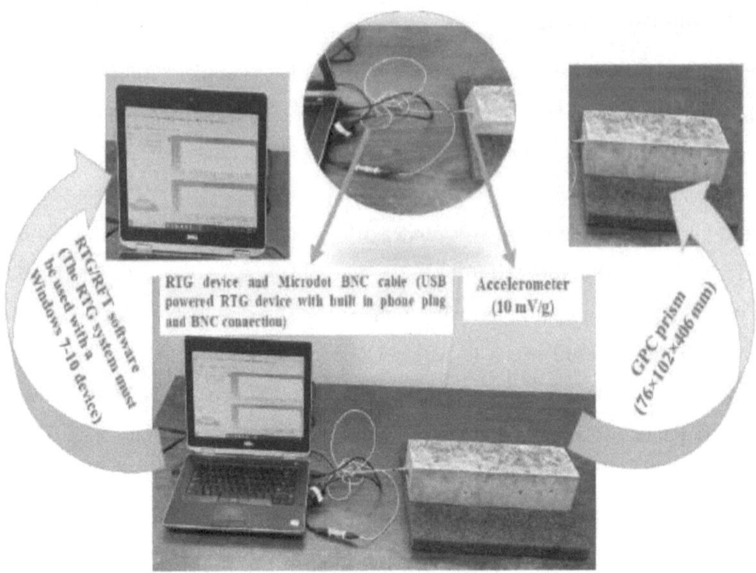


**Figure 19. RFT/RTG Device**

### 6.1.5.4   Freeze-thaw damage model

Hongfa et al. [181] used Aas-Jakobsen's [187] S-N equation (Equation 13) and proposed a model (Equation 14) for calculating the freeze-thaw damage of cement-based concrete in both the laboratory and real environmental conditions. According to their assumption, the freeze-thaw fatigue damage in concrete can be calculated as Equation 14:

$$S = \frac{\sigma_{max}}{f_t} = 1 - \beta\,(1 - R)\,\log N \tag{13}$$

$$D_n = 1 - \frac{0.6\,\log N}{\log N - 0.4\,\log(N-n)} = 1 - \frac{E_n}{E_0} \tag{14}$$

where:

$D_n$ = freeze-thaw damage variable expressed as the RDME.

N = fatigue life of concrete exposed to freeze-thaw cycles.

n = number of freeze-thaw cycles.

$E_0$ = initial dynamic modulus of elasticity.

$E_n$ = dynamic modulus of elasticity after 'n' freeze-thaw cycles.

$\beta$ = concrete material parameter

They also determined the value of $\beta$ under freeze-thaw conditions using Equation 15 and established a relationship between $\beta$ and compressive strength.

$$\beta = \frac{Dn}{\log N} \tag{15}$$

In this study, the value of $\beta$ of fly-ash based GPC and bottom-ash based GPC and other by-products-based GPC performed by Zhao et al. [40] and Mengxuan et al. [182] was calculated to find the applicability of Equation 15. Table 10 compares the mix proportions of other researchers [40,182] to the ones used in this study. Equation 14 and Equation 15 were used to calculate the value of $\beta$ of each mix proportion (shown in Table 10), and to establish the relationship between $\beta$, number of cycles and compressive strength of each mix proportion.

### 6.1.5.5 TCLP test

Several materials are categorized as hazardous waste when they are dumped in landfills. GPC is considered as a toxic material due to the use of hazardous waste materials and chemical activators in its mixture. This is why, it is vital to check its chemical metals leach-ability. In this study, the TCLP test was used to analyze the leaching of chemical metals from GPC. To study the environmental compatibility of K-based GPC made by 50% fly-ash and 50% bottom-ash, TCLP was performed as per USEPA 1311.

In this study, GPC was crushed to achieve a minimum of 100 g to characterize the chemical metals. Then, 5 g of GPC sample was mixed with 95 mL of distilled water to measure its pH level. After the determination of pH value, a proper extraction fluid was selected and was added to $100 \text{ g} \pm 0.1 \text{ g}$ of GPC sample (< 9.5 mm). The solution was placed in the extraction vessel and was rotated at $30 \pm 2$ rpm for $18 \pm 2$ hours at ambient temperature ($23 \pm 2$ °C). At the end of the extraction period, the solution was transferred to the filter holder. The filtrate collected is called leachate. According to USEPA 1311, 100 grams of a GPC sample can provide proper values of chemical metals leach-ability. Moreover, the leaching values of this GPC sample was compared with other literature and is in good-agreement with other studies.

Table 10. Comparison of mix proportions

| Sample | Mixture | Fly-ash | Slag | Cement | Sand | Gavel | Na$_2$SiO$_3$ | NaOH | Water | Water reducer | Curing | Si/Al | L/S |
|---|---|---|---|---|---|---|---|---|---|---|---|---|---|
| 1 | PCC [40] | 56 (kg/m³) | 56 (kg/m³) | 448 (kg/m³) | 626 (kg/m³) | 1022 (kg/m³) | - | - | 174.7 (kg/m³) | 2.5 (kg/m³) | Standard | - | - |
| 2 | GPC-10 [40] | 346.7 (kg/m³) | 38.5 (kg/m³) | - | 601.7 (kg/m³) | 1203.5 (kg/m³) | 165.7 (kg/m³) | 66.2 (kg/m³) | - | - | 80 ºC | - | - |
| 3 | GPC-30 [40] | 269.6 (kg/m³) | 115.6 (kg/m³) | - | 601.7 (kg/m³) | 1203.5 (kg/m³) | 165.7 (kg/m³) | 66.2 (kg/m³) | - | - | Standard | - | - |
| 4 | GPC-50 [40] | 192.6 (kg/m³) | 192.6 (kg/m³) | - | 601.7 (kg/m³) | 1203.5 (kg/m³) | 165.7 (kg/m³) | 66.2 (kg/m³) | - | - | Standard | - | - |
| 5 | RMSFFA-SA2.0-NA0.6-RT-28D [182] | - | - | - | - | - | - | - | 25% | 25% | Room temp. (23ºC) | 2.00 mol | 0.53 |

| Sample Mixture | 6 | 7 | 8 |
|---|---|---|---|
| | RMSFFA-SA2.0-NA0.6-80C-28D [182] | Fly-ash based GPC (current study) | Bottom-ash based GPC (current study) |
| Fly-ash | - | | |
| Slag | - | | |
| Cement | - | | |
| Sand | | | |
| Gavel | - | Available in Table 9 | Available in Table 9 |
| $Na_2SiO_3$ | - | | |
| NaOH | - | | |
| Water | 25% | | |
| Water reducer | 25% | | |
| Curing | 80ºC | | |
| Si/Al | 2.00 mol | | |
| L/S | 0.53 | | |

### 6.1.6 Results and discussion

#### 6.1.6.1 Physical characteristics

The slump test was performed for the analysis of viscosity behavior of fly-ash based GPC and bottom-ash based GPC to investigate the workability, according to ASTM C143 [188]. The average slump value of fly-ash based GPC and bottom-ash based GPC was measured as 245 mm and 215 mm, respectively. The bottom-ash based GPC specimens showed lower workability than fly-ash based GPC specimens because according to the microscopic study of GPC, the smooth surface and rounded-shape of the fly-ash particles improve ball-bearing effect, which increase the workability [189]. Moreover, it can be seen from Table 9 that bottom-ash has a lower specific gravity when compared to fly-ash. Hence, when the same weight of fly-ash per cubic meter of the material is replaced with bottom-ash, there is more dry volume of the material, which may also be partly responsible for reducing the workability of the mix.

The average dry density of fly-ash based GPC and bottom-ash based GPC at 7 days was 2415 kg/m$^3$ and 2422 kg/m$^3$, respectively. The dry density of these two types of GPC increased as the age of the GPCs increased. The average dry density of fly-ash based GPC increased from 2415 kg/m$^3$ to 2431 kg/m$^3$ when age of samples increased from 7 days to 28 days with an overall increase of 0.66%. In contrast, the average dry density of bottom-ash based GPC increased from 2422 kg/m$^3$ at 7 days to 2435 kg/m$^3$ at 28 days with an overall increase of 0.53%. The authors attribute this slight difference in density to in-batch test variability.

It is well-known that the specific gravity of fly-ash and bottom-ash is comparable because these two materials have similar chemical compositions. Table 9 indicates that fly-ash has higher specific gravity compared to its counterpart bottom-ash. It is reported that cenospheres and poor gradation of particles degrade specific gravity of bottom-ash [190].

#### 6.1.6.2 Compressive strength of GPC

It is well-known that elevated curing temperature and duration are beneficial toward the acceleration of the polycondensation process. In previous studies performed by current authors, three methods of curing (ambient, steam and dry curing) were used to achieve higher compressive strength. A minimum of six specimens (100 × 200 mm) were cured at ambient, 30, 45, 60, 80 °C for 24 hours and then kept at room temperature for 28 days. These samples were tested at the age of 28 days by using Forney compressive testing machine model #AD 650. However, steam-cured GPC at a temperature of 80 °C for 24 hours following by 28 days of room temperature curing achieved higher average compressive strength (35 MPa). Based on the other studies on the microstructure of GPC [191], the steam curing method improves the dissolution rate of chemical species, such as Silicon Dioxide ($SiO_2$) and Aluminum Oxide ($Al_2O_3$), from mixture where the rate of geopolymerization increases. This finding can be attributed to the full and uniform internal curing of specimens. This finding is also in good-agreement with Yewale et al. [192], where the optimum compressive strength of steam-cured GPC was achieved at 80 °C. So, in the present study, GPC samples were steam-cured at 80 °C for subsequent experiments. However, it should be mentioned that GPC samples can be steam-cured at 60 °C for some applications and it may not be necessary to cure at high temperature such as 80 °C.

The same method and temperature (80 °C) were then used to produce fly-ash based GPC. A minimum of three fly-ash based GPC cylinders were cast to find their average compressive strength at the age of 28 days. The compressive strength of fly-ash based GPC specimens was 32 MPa, 30 MPa and 32 MPa with an average of 31 MPa. It can be seen from the results of the compression test that the compressive strength of bottom-ash based GPC steam-cured at 80 °C is higher than fly-ash based GPC with a similar curing regime. Although several studies have reported that bottom-ash based GPC has lower compressive strength compared to fly-ash based GPC [25,69,71], it can be attributed to the higher porosity of bottom-ash [193]. However, it should be noted that the ratio of $SiO_2/Al_2O_3$ also affects the compressive strength of GPC [194,195]. This means that GPC made by waste ashes with a higher of $SiO_2/Al_2O_3$ ratio tend to develop higher compressive strength. In this study, the bottom-ash had a higher $SiO_2/Al_2O_3$ ratio than fly-ash. This is why bottom-ash based GPC had higher compressive strength compared to fly-ash based GPC.

Generally, Na-based solution is mostly considered in various studies due to its low cost, availability, desired workability and durability. However, the K-based solution can also be used for high temperature applications [132,196,197]. According to Hounsi et al. [133], Na-based GPC gains lower compressive strength value than K-based GPC at the same alkali concentration of the current study (12 M). Hounsi et al. [133] attributed this phenomenon to the reduction of Si/Na ratio at a high concentration of NaOH, where NaOH slows the polycondensation process and reduces the mechanical properties of Na-based GPC.

In previous studies performed by current authors, attempts have been made to cure the samples at ambient temperature. However, as aforementioned, the higher compressive strength is achieved at 80 °C. It is well known that only ambient temperature curing is a practical method in the construction field for GPC and to save energy. Hence, mixing calcium-based material such as cement with ashes is suggested to improve setting time, workability and durability of GPC cured at ambient temperature [198]. The microstructural investigation of fly-ash based GPC mixed with cement showed that the geopolymerization process is more likely as calcium alumino-silicate hydrate (C-A-S-H), which contributes to hardening and early strength gain of GPC mix with cement. Moreover, the enhanced strength of GPC mixed with cement is attributed to the generated heat during the geopolymerization process, where cement helps GPC to initiate condensation reaction at ambient temperature [186].

### 6.1.6.3   Freeze-thaw resistance

### 6.1.6.3.1   Mass loss of specimens during the freeze-thaw process

The weight of GPC was taken every 30 cycles of freeze-thaw to calculate their mass loss. Figure 20 shows the average mass loss of six fly-ash based GPC and six bottom-ash based GPC up to 300 cycles of freeze-thaw. Fly-ash based GPC shows higher and rapid mass loss possibly because the specimens' structure failed at the early age of freeze-thaw cycles, and due to the poor bonding in the Interfacial Transition Zone (ITZ) which led to severe surface scaling. This finding is in good-agreement with the mass loss result of fly-ash based GPC-10 studied by Zhao et al. [40]. In contrast to fly-ash based GPC, the mass loss of bottom-ash based GPC was slow till 300 freeze-thaw cycles which indicates higher bonding strength of paste of bottom-ash based GPC.

### 6.1.6.3.2   RDME of fly-ash based and bottom-ash based GPC

Figure 21 indicates the average RDME reduction of six fly-ash based GPC and six bottom-ash based GPC over 300 cycles of freeze-thaw. As can be seen in Figure 21, bottom-ash based GPC exhibits higher RDME than fly-ash based GPC. The RDME of both types of GPCs dramatically dropped when cycles increased from 0 to 60. Authors attribute this RDME reduction to the existence of uncured by-product particles in the microstructure of GPC that caused GPC specimens to lose its microstructural strength at early freeze-thaw cycles [98].

It also can be seen that RDME loss of bottom-ash based GPC (13.4%) was higher than fly-ash based GPC (10.9%) until 60 cycles. However, RDME of fly-ash based GPC reduced intensely until 150 cycles (RDME ≈ 60%). In this study, the freeze-thaw test for fly-ash based GPC was continued up to 300 cycles, even though according to ASTM C666 [9], the freeze-thaw testing procedure of specimen must be stopped when its RDME reaches 60% of the initial modulus. Since concrete is a heterogeneous material, the issue mentioned above might be due to the various factors such as the density and RDME of the main constituents (such as fly-ash and bottom-ash) and the characteristic of the ITZ which affect the elastic behavior of the composite [79]. Moreover, in general, the RDME of the fly-ash based matrix (durability factor of 27) and bottom-ash matrix (durability factor of 48) are determined by their porosity. So, the parameters determining the porosity of the matrix, such as geopolymerization process, curing conditions, AEA amount, etc., could be the other reasons for rapid RDME reduction of fly-ash based GPC.

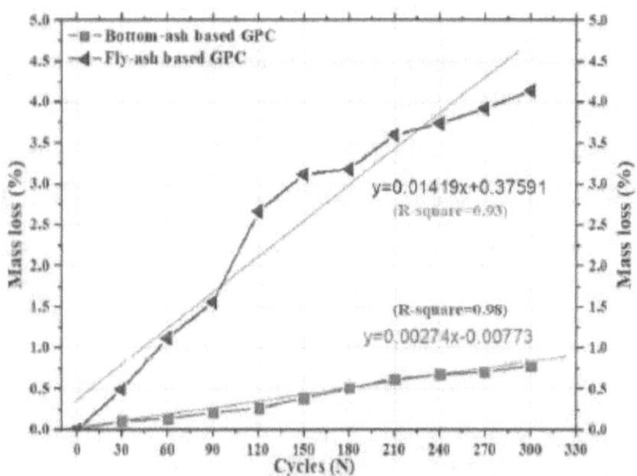


**Figure 20. Average mass loss of fly-ash based GPC and bottom-ash based GPC over 300 cycles of freeze-thaw**

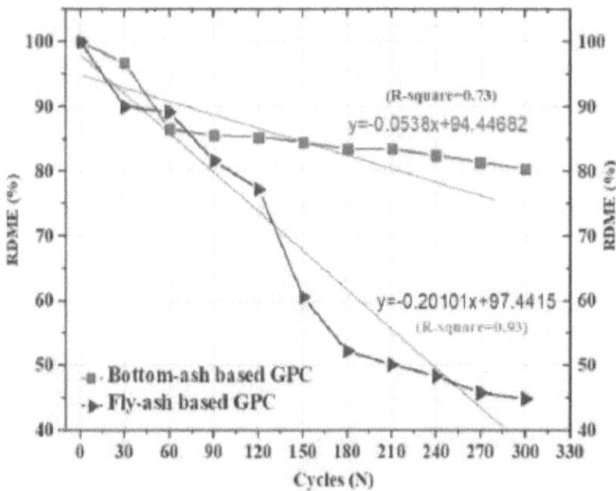


**Figure 21. Average RDME reduction of fly-ash based GPC and bottom-ash based GPC vs. No. of cycles**

It is well-known that AEA provides more free spaces for the water to freeze. However, the pressure made by cryo (ice formation inside the air voids of GPC) can overcome the tensile strength of the matrix when these free spaces are overloaded with cryo. Consequently, this pressure creates micro-cracks in the ITZ and would reduce the ultimate strength of concrete. So, fly-ash based GPC can bear lower force when exposed to the freezing and thawing conditions since the initial compressive strength of fly-ash based GPC (~32 MPa) was lower than bottom-ash based GPC (~35 MPa).

### 6.1.6.4  Leach-ability of GPC
### 6.1.6.4.1  Laboratory investigation of bottom-ash based GPC

Identification of heavy metals and their leaching is one of the important factors that govern the utilization of GPC because of its influence on the environment. The toxicity of heavy metals depends on their concentration rate in the environment. With the increasing concentration rate of heavy metals in the environment, these toxic hazardous metals can be accumulated in living tissues and cause irreparable events. So, the TCLP test was conducted for obtaining the essential data about main chemical elements such as Si, Al, Na and other hazardous metals such as Cr, Cu, Hg ,etc. It should be mentioned that the TCLP test was only possible on one sample due to the cost prohibitive nature of this test. Bottom-ash based GPC was selected in this study because of its higher freeze-thaw resistance than fly-ash based GPC. The TCLP test was performed by the Maxxam Analytics lab, Victoria, Canada. Table 11 shows the pH of bottom-ash based GPC at three levels of extractions. The initial pH of the sample was 11.6 due to the existence of alkali constituents (K-based) in the bottom-ash based GPC mixture. To prepare the leaching/extraction fluid, 5.7 mL glacial acetic acid ($CH_3CH_2OOH$) was added to 500 mL reagent/distilled water. When reasonably mixed, the pH of this fluid was 4.96. Then, to perform the TCLP test, an amount of the proper leaching fluid equivalent to 20 times the mass of the specimen (20:1 liquid to solid ratio) was added to the extraction vessel of TCLP equipment. Then, the measured pH of leachate was 6.36.

Table 11.  pH values of bottom-ash based GPC

| pH | - |
|---|---|
| Initial pH of sample | 11.6 |
| Final pH of leachate | 6.36 |
| pH of leaching fluid | 4.96 |

The results of the TCLP test showed that all the heavy metals could be effectively immobilized into the geopolymeric paste. Figure 22 indicates that Ba, B and Fe have the highest leaching concentration and the rest of heavy metals have a concentration of less than 0.10 mg/L. Moreover, the obtained results showed that the concentration of all the heavy metals is below the regulatory level in accordance with USEPA 1311 and USEPA CFR. This could be attributed to the cations of heavy metals (such as $Cu^{2+}$, $Cd^{2+}$, $Fe^{3+}$, $Zn^{2+}$, $Pb^{2+}$, and total Cr) that can participate in the balance of the negative charge of tetra-silicate ($[SiO_4]^{4-}$) and potassium tetra-aluminate ($[K^+AlO_4]^{4-}$) [199,200]. So, bottom-ash based GPC showed low porosity, which could help immobilize all the heavy metals [201,202].

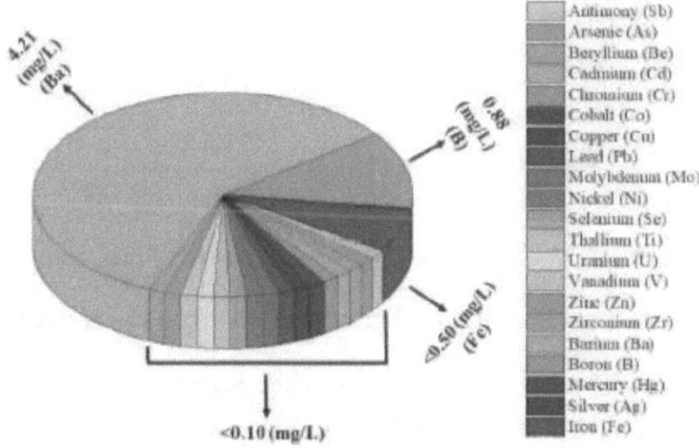


**Figure 22. TCLP test results of bottom-ash based GPC**

### 6.1.6.4.2 Observational study on leach-ability of GPC during the freeze-thaw process

In this study, the leach-ability of fly-ash-based GPC and bottom-ash based GPC was measured using a water testing method called HACH strips. The water sample was collected every 30 cycles from the freeze-thaw cabinet to measure the leach-ability of fly-ash based GPC and bottom-ash based GPC. Since the trend of data for every 30 cycles was quite similar only results obtained from the last cycle (300 cycles) were analyzed and plotted in Figure 23. The leach-ability of both fly-ash based GPC and bottom-ash based GPC was constant over 300 cycles of freeze-thaw (Figure 23). This phenomenon shows that the geopolymerization process was fully completed, and all toxic metals were trapped in the paste. Moreover, heat-treatment improved the microstructure of the GPC specimens, decreased the porosity and decreased the leach-ability of the paste [203].

It also can be seen that the concentration of all the elements is almost in same range. However, fly-ash based GPC had more amount of alkalinity than bottom-ash based GPC, possibly due to poor geopolymerization of fly-ash based GPC compared to bottom-ash based GPC, which led to leaching of potassium from fly-ash based GPC.

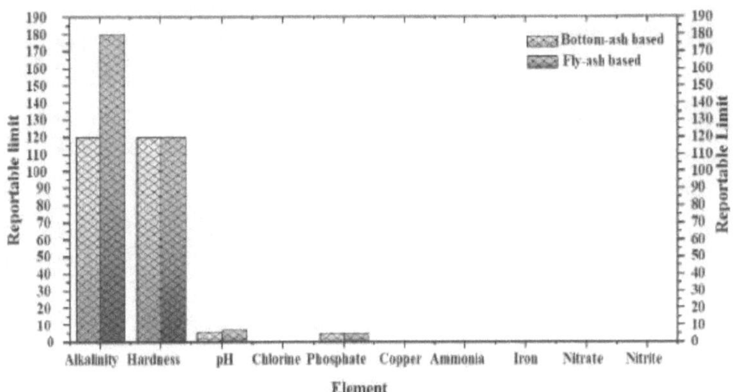


**Figure 23. Leaching range of GPCs over 300 freeze-thaw cycles**

### 6.1.6.5 The Relationship between β, number of cycles and compressive strength of GPC

Figure 24 shows calculated values of β for data obtained from Zhao et al. [40] and Mengxuan et al. [182] and compares with values of β of fly-ash based GPC and bottom-ash based GPC. The value of β for mix #1, 2, 3, 4, 5, 6, 7 and 8 is 0.19, 0.44, 0.22, 0.19, 0.27, 0.23, 0.18 and 0.16, respectively. According to Equation 15, the value of β is inversely proportional to the logarithmic function of N, which means a higher number of freeze-thaw cycles give a lower value of β. The results of the current experiment, shown in Figure 24, confirm the finding mentioned above that bottom-ash based GPC (mix #8) with a lower value of β has higher freeze-thaw resistance than fly-ash based GPC (mix #7). This finding is also applicable to the rest of the GPC samples. Moreover, it can be seen that the value of β for GPC samples with compressive strength less than 50 MPa decreased abruptly. While, β value of GPC samples with compressive strength higher than 50 MPa reduced gradually. Although the results of the current study are in good agreement with Hongfa et al. [181], the authors suggest that compressive strength of GPC should be considered in Hongfa et al.'s equation [181] because compressive strength is one of the key properties of GPC revealing internal bonding of paste. Therefore, a greater number of freeze-thaw cycles were achieved when GPC had higher bonding strength. This finding can be seen in Figure 24 where the value of β was varied when compressive strength increased (the value of compressive strength of mix #1, 3, 4 and 8 was higher than their β, while the compressive strength of mix #2, 5, 6 and 7 was lower than their β).

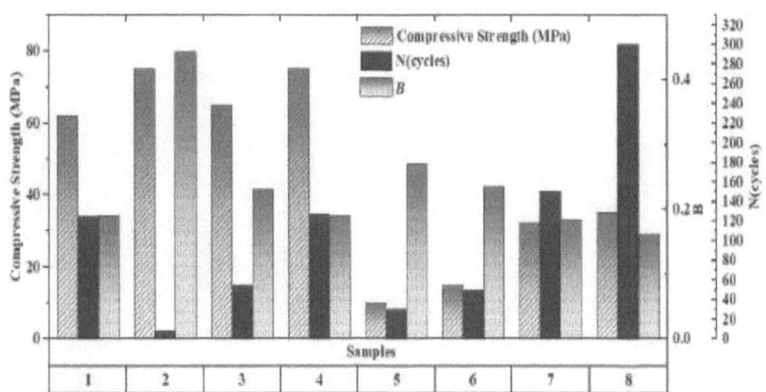

Figure 24. Value of β, compressive strength and number of cycles for different types of GPC

## 6.1.7  Conclusion

In this study, the deterioration and mass loss produced in fly-ash based GPC and bottom-ash based GPC during 300 cycles of freeze-thaw were evaluated using RFT. The leach-ability of bottom-ash based GPC was also measured using the TCLP test to characterize the heavy metals. According to the obtained results:

- Bottom-ash based GPC indicated lower mass loss than fly-ash based GPC. Authors attributed the phenomenon mentioned above to the poor bonding of pastes in the ITZ in fly-ash based GPC.

- The resonant frequency of both types of GPC was measured after exposure to 300 freeze-thaw cycles with interval of 30 cycles. According to the results, bottom-ash based GPC showed better freeze-thaw resistance than fly-ash based GPC. It could be attributed to various parameters including geopolymerization process, curing conditions and the amount of AEA.

- Toxicity of heavy metals leaching from bottom-ash based GPC was measured using the TCLP test. The results showed that all the heavy metals including Si, Al, Na, Cr, Cu, Hg etc. were trapped and immobilized in the paste, and all of them were below the standard range of USEPA 1311 and USEPA CFR.

- A comparison between compressive strength, N and $\beta$ of different types of by-products-based GPC was made. The experimental results showed that a higher number of freeze-thaw cycles give lower $\beta$. So, Bottom-ash based GPC with a higher number of cycles had lower $\beta$ (0.1614) than fly-ash based GPC (0.1838). The authors also found that compressive strength should be accounted for in the proposed equation by Hongfa et al. [181].

## Chapter 7   Investigating Corrosion Resistance of GPC

The microstructure of GPC and conventional concrete is different in terms of their chemical and pores characteristics, particularly due to the high alkalinity of GPC. Thus, differences in corrosion mechanisms of steel elements embedded in these types of concrete could be expected, with significant implications for the durability of reinforced concrete elements. This study assesses the corrosion behaviour of steel embedded in GPC and conventional concrete exposed to aggressive conditions. In this regard, in this Chapter, corrosion of PCC and GPC was measured using half-cell potential and linear polarization resistance methods.

The following paper has been published in the journal of "**Substantiality**". It should be noted that this chapter is an edited version of the published manuscript.

## 7.1 Introduction

Green economy, biocarbon economy and low-carbon economy are some of the terms that indicate the global efforts for decarbonisation [204]. The onus is on the mitigation policies for reducing greenhouse gas emissions [205]. The year 2017 recorded the highest inclination of the order of 32.5 Gt. $CO_2$ emissions [206] out of which 39-28 % is from building and construction sector and 11 % is from building materials and transport activities. The construction industry being a carbon-intensive sector, consuming significant amounts of energy, products and services of different sectors, is significantly challenging sustainable growth [207,208]. In the entire spectrum of construction industry, the production of cement alone produces the largest amount of $CO_2$ and is the $2^{nd}$ largest source of $CO_2$ emission worldwide. Henceforth, global environmental impact of cement production has resulted in increased momentum of research for other alternatives. The utilization of Supplementary Cementitious Materials (SCM's) such as fly-ash, rice husk ash, slag etc. as partial replacement of cement are being extensively explored. Their potential for enhancing the binder characteristics and reduction of environmental impact has been well recognized [209].

The exploration of solution for a two-faced problem i.e., reduction of $CO_2$ emissions and the industrial waste management, has led to the development of an alkali activated concrete known as GPC. The term 'geopolymer' was first used in Davidovits work of the formation of polymeric Si-O-Al bonds from chemical reaction of alkali silicates with aluminosilicate precursors [210]. As per the Duxson model [211], the process of geopolymerization involves three steps: 1) the dissolution of aluminosilicate materials and the release of silicate and aluminate monomers$[Si(OH)_4]$- and $[Al(OH)_4]$-; 2) Initial gels(mono cross linked systems) are being produced by co-sharing of oxygen atoms from the reactive silicate and aluminate monomers, the process being known as condensation; 3) In the last stage the initial gels are converted into the geopolymers gels and the process is knows as polycondensation. Different industrial waste materials such as fly-ash, metallurgical slag, metakaolin, mining wastes, silica fume etc. could be used as source materials for geopolymerization. Although, the reactivity depends upon their physical, chemical and morphological properties but a stable geopolymer requires the source material to possess the following charcateristis:1) Highly amorphous; 2) enough reactive glassy content; 3) low-water demand; 4) ability to release aluminium easily. A more detailed model for geopolymerization given by Provis et al. [212], discusses the synthesis of geopolymer and zeolites by the geopolymerization of metakaolin and fly-ash. The model explores the different silicate oligomers for their inclusion in alkali solution. The respective oligomers polymerise into geopolymer fragment and aluminosilicate nuclei. Finally, the geopolymer gels and zeolites are formed by the polycondensation of the remaining silicate monomers. Both microstructural and the chemical properties of geopolymers with different source materials, will vary greatly despite their physical properties might appear to be similar. Alkali activators, an important constituent of geopolymerization, are used to activate the aluminosilicates materials. The most commonly used alkali activators are NaOH, KOH, $Na_2SiO_3$ and $K_2SiO_3$. Although KOH induces higher alkalinity in the mix than NaOH, but the capacity to liberate higher amount of silicates and aluminate monomers makes NaOH most suitable.

After lime and Portland cement, geopolymer could be considered as the third-generation cement. Geopolymers can be modified by correctly selecting the raw materials and optimising the design mix. Similar to PCC, aggregates can be added into geopolymer to produce GPC. The characteristics of GPC has been reported to be better than normal concrete. Since the input materials for the geopolymer mix could be different, therefore the final products of hydration are different from those produced in the hydration of cement [213][214]. The liquid-solid ratio, $SiO_2/Al_2O_3$, $R_2O/Al_2O_3$, $SiO_2/R_2O$ ratio majorly impacts the properties of geopolymer pastes. Numerous authors have suggested that an amorphous structure of geopolymer will result better mechanical properties [214- 217]. An increase in alkali content and corresponding decrease in silicate content results in the formation of aluminosilicate network structures, increasing the compressive strength of the mix. The development of widely distributed and larger sized pores in a fly-ash based paste made with a very small dosage (18%) of activators have been reported. On the other hand, a narrow distribution and a smaller sized pore were seen with a higher activator dosage of the order of 30%. This is primarily due to the refinement of pores by the dissolution of particles and the formation of products. The reduced porosity improved the strength of the paste [217]. Apart from the changes in the strength, the increase in the ratio of $SiO_2/Al_2O_3$ increases the setting time of the paste as well.

Along with mix-ingredients, curing conditions greatly impact the characteristics of GPC [22,32, 61,62,87,218]. Curing of GPC can be done in three ways: 1) Heat curing; 2) Steam Curing; 3) Ambient Curing. The curing temperatures required for consummation of geopolymerization ranges between $40\,^0C$-$85\,^0C$. In a study, it has been reported that the compressive strength a fly-ash based GPC, was much higher when cured at a temperature of $85^0C$ than at lower temperatures [22]. However, the ascent in compressive strength significantly reduced after 24 hours of casting. Further, for a metakaolin based GPC, curing at 30 % Relative Humidity (RH) was preferred than at 70 % for better results [219]. In another study, it was concluded that curing of metakaolin based GPC should be done at higher temperatures ($40^0C$-$100^0C$) for significant strength gain [87]. In contrast to that, it was also suggested that thermolysis of Si-O-Al-O bond would result in failure of samples, if high temperature curing is continued for longer durations. For achieving the desirable mechanical and durability characteristics, adequate curing of GPC is required [218].

Although the strength of matrix directly depends upon the microstructural and nano-structure of materials used for its production, but it greatly influence the permeability and porosity of the structure. The formation of geopolymers from geopolymeric aluminosilicate hydrate gel (A-S-H) differentiates it from normal concrete made from calcium silicate hydrate gel (C-S-H). Further, the low calcium content of GPC from that of normal concrete reduces the durability issues significantly. Hence, a differential durability behaviour (design life & performance) of GPC structures should be expected than that of normal concrete. Steel corrosion is one of the major durability issue that has impacted the long term performance of structures made with normal concrete. Chloride generates corrosion by damaging the passive film of steel reinforced rebar [220]. Some reports stated [221] that improving the quality of GPC can prevent the corrosion of reinforcement bar. Tennakoon et al. [222] performed long term tests on corrosion of steel rebar in fly-ash based GPC and slag-based GPC. The results of their study showed that chloride diffusion coefficient is less in fly-ash and slag GPC than that of conventional concrete. With the increase of slag content in the binder, the diffusion coefficient decreased. Tennakoon et al. [222] have also concluded that the embedded rebar in fly-ash and slag based GPC has higher resistance to corrosion than a rebar in PCC. Reddy et al. [223] conducted an experiment on durability of reinforced GPC in the marine environment. Reddy et al. [224] evaluated the corrosion based durability of low calcium fly-ash based GPC using beams that are centrally reinforced, made with 8M and 14M concentrations Na-based solutions. The experimental results proved that GPC has better corrosion resistance performance compared to PCC. Regarding the corrosion of reinforcement in GPC, the literature is relatively limited; there are a few work reported on corrosion resistance evaluation of steel rebar in fly-ash based GPC [186,225]. Chindaprasirt et al. [226] investigated the effect of NaOH molarities on chloride penetration, steel corrosion and compressive strength of fly-ash based GPC under marine condition. The concentrations of NaOH of 8, 10, 12, 14, 16 and 18 molars were used. The specimens were cured at ambient temperature in laboratory for 28 days and then were exposed to marine condition. The specimens were tested for compressive strength, chloride penetration and corrosion of embedded steel bar after three years of exposure to marine conditions. According to the results, both chloride penetration and corrosion of embedded steel reduced with the increasing of NaOH concentration. The corrosion is high with the GPC of low compressive strength. In addition, increasing the NaOH concentration in GPC resulted in a decrease in the chloride binding capacity. The rapid chloride permeability test (RCPT) ASTM C1202 [12] provides information on the instantaneous chloride diffusion of concrete as it is conducted within a short period of time (6 hours). Lee et al. [227] investigated RCPT of fly-ash and slag based GPC activated by sodium-based solution. According to the RCPT results, the total charge passed through GPC samples decreased to 759 coulombs after cured for 270 days. This is attributed to denser geopolymer paste and the reduced penetrability of the chloride ion due to the longer curing period. Chandani et al. [228] investigated RCPT of PCC and GPC made by fly-ash and bottom-ash. The RCPT of samples was performed at 28 and 180 days and total charges passed through the concretes was measured. According to the results, the charges passed through 28 days GPC is in the range of 1000–2000 C which is categorized as low permeable concrete. Although, PCC has slightly higher value than the corresponding value for GPC. So, 28 days PCC can be also categorized as low permeable concrete. Moreover, at 180 days, the charges passed through GPC are lower than in 28 days. This changes the permeability category of GPC to "very low" permeable concrete range. Furthermore, All the GPC samples achieved the same level of permeability level after 365 days, with charges passed through GPC being 300–330 Columb.

From safety perspective, it is essential to perform routine assessments of concrete to identify the damages in the areas with critical levels of corrosion [220]. There are many destructive methods for corrosion assessment of GPC structure, which requiring material samples to be taken from structure. In this regard, NDT methods must be developed for testing without destructing the structure. The only standardized test methods for corrosion monitoring is the ASTM C876 [229], half-cell potential techniques [229]. Hence, measurement of steel corrosion in GPC using two NDT methods, Half Cell Potential (HCP) and Linear Polarization Resistance (LPR), can be an effective indicator of steel reinforcement corrosion rate. Gunasekara et al. [230] used HCP and LPR methods to experimentally study the corrosion rate of steel rebar in GPC produced from three types of fly-ashes over 540 days of exposure to 3% sodium chloride (NaCl) solution. The corrosion rate values for these GPCs ranged from 0.55 to 1.65 $\mu A/cm^2$ where these findings were almost equal to conventional concrete (0.65-1.20 $\mu A/cm^2$). In another study, Miranda et al. [231] that HCP values for the passivated reinforcement in GPC (ranging between -100 to -200 mV) is moderately similar to the PCC mortar results.

The focus of this paper is to develop a novel fly-ash based GPC with 50 % replacement of bottom-ash. Further, in meso-scale level, the durability behaviour of the matrix in terms of resistance to chloride–induced corrosion in comparison with ordinary concrete is also investigated. Two NDTs i.e., HCP and LPR, have been used to study the corrosion behaviour.

## 7.2 Experimental program

### 7.2.1 Precursor materials

The properties of precursors have been already explained in section 4.5.1.

### 7.2.2 Alkali activator

The properties of alkaline activators have been already explained in section 4.5.2.

### 7.2.3 Portland cement

Type 1 Portland cement was used for the manufacture of the control mix concrete. The control concrete mix was designed to have a target strength of 35 MPa similar to that of GPC.

### 7.2.4 Aggregates

The properties of aggregates have been explained in section 3.5.2.

## 7.3 Method of casting and curing

### 7.3.1 Specimen dimensions

In this study, mix proportion and curing method of GPC build on previous work done in the Facility for Innovative Materials and Infrastructure Monitoring (FIMIM) by the authors [232- 234]. The dimension of beams were $6''\times6''\times21''$ ($152\times152\times533$ mm), centrally reinforced with $\frac{1}{2}''$. The rebar was about $\frac{1}{2}''$-$1''$ outside the beam to facilitate connecting the wire for testing purpose.

### 7.3.2 Specimen preparation

The production of GPC samples has been already explained in section 4.6.1 and 4.6.2. The mix proportions of GPC beams (obtained from previous work performed by the authors) and PCC beams are already shown in Table 4 and Table 12, respectively.

In order to prepare the PCC beams, the dry ingredients such as cement, sand and coarse aggregates were mixed for about 2 minutes and the water was added slowly until the concrete was uniform and workable. After compaction on a vibrator table, each beam was labelled and then covered with a plastic sheet to prevent evaporation and kept overnight in the laboratory at ambient temperature. The PCC samples were demolded and stored in a water tank at $23\pm2$ °C for 28 days. Further, #4 (imperial units) steel rebar of nominal diameter 12.5 mm and length 650 mm was embedded in concrete in such a way that 100 mm of the total length remained exposed. Commercially available hot rolled steel reinforcing bars with a yield strength of 400 MPa were used. The rebar was protruded so that a connection could be made for impressed current corrosion. The embedded rebar was corrosion free during casting. After rolling, the rebar was kept in oil. Thereafter, acetone was used to clean the rebar of oil before embedding it into concrete. The rebar was weighed to the accuracy of 1 gram. In order to prevent the crevice corrosion at the steel concrete interface and to serve as a bond breaker, the rebar was wrapped with Teflon tape. The protruded 100 mm length of the rebar was epoxy coated to protect from the corrosive environment.

**Table 12. Mix proportion of PCC**

| Material | Content (kg/m³) |
| --- | --- |
| Cement | 400 |
| Sand | 660 |
| 12.5mm aggregate | 701 |
| 6.5mm aggregate | 467 |
| Water | 160 |

## 7.4    Methodology

After 28 days, when GPC and PCC samples cured in ambient temperature (followed by 24 hours steam curing) and water tank respectively, beams were placed in a chloride solution for a period of 28 days. This helps to keep the initial D.C. power to a manageably low value. Figure 25 is a schematic of accelerated corrosion test setup in 3.5% NaCl solution used in this work. As can be seen in Figure 25, the corrosion tank was filled with chloride solution that would allow each beam to be partially immersed. After that, the extended side of rebar acting as an anode was connected to a 30V power supply. A stainless steel rod was used to act as a cathode. The D.C. power supply was turned on and set to 30V electrical potential. This voltage was chosen based on work conducted by Reddy et al. [224] to make the steel rebar as anodes, accelerate the corrosion process and decrease the test period time.

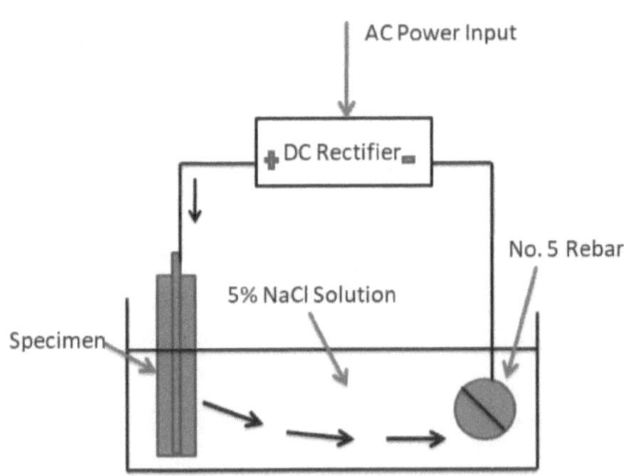


**Figure 25. Schematic of accelerated corrosion test setup**

Once the testing procedure started, current readings were taken every 24 hours. A rise in the current indicated the beginning of corrosion process, and eventually the start of the formation of cracks in the beams. Once the beams reached a high current value, there were visible signs of corrosion and cracking of beams and the beams were considered to be failed. The time taken to initiate the corrosion in the rebar in GPC was higher than that of PCC concrete. The beams were removed from the chloride water and left to air dry for 24 hours. After that, the beams were tested for rate of corrosion using LPR. The final step involved splitting the PCC and GPC beams to recover the rebar and to determine the effects of corrosion on rebar by measuring the mass loss of the steel rebar in each beam.

### 7.4.1   HCP

HCP is an effective method that has been used by many researchers across the world [235] and ASTMC 876 [229] gives the guideline for predicting corrosion activity (Table 13). It is a method of assessing invisible corrosion of reinforced concrete without destructing the samples. HCP provides information about probability of corrosion. Table 13 gives the guide for evaluation of corrosion activity versus a standard copper/copper sulphate half-cell.

**Table 13. Predicting corrosion using Cu/CuSO4 half-cell**

| Corrosion Risk | Half-cell potential (versus Cu/CuSO$_4$) |
| --- | --- |
| **Severe Corrosion** | Less than -500 mV |
| **High Corrosion Risk (90% probability)** | Between -500 mV & -350 mV |
| **Medium corrosion risk (50 % probability)** | Between -350 mV & -200 mV |
| **Low Corrosion Risk (10% probability)** | Higher than -200 mV |

Figure 26 indicates schematic of the HCP measurement setup. In accordance with ASTM C876 [229], a digital voltmeter is used to read the potential difference values between the external reference electrode and reinforced steel rebar. In this study, copper/copper sulphate was used as a reference electrode. If the surface of the concrete is too dry, pre-wetting is required. A pre wetted sponge is used to ensure proper surface contact between the concrete surface and the tip of half-cell electrode.

For a consistent reading, a center line with a pre-defined equal spacing of three measuring points at 7 inches distance was marked on the surface of the concrete. The potential values for these three points were recorded from the voltmeter for both PCC and GPC beams.

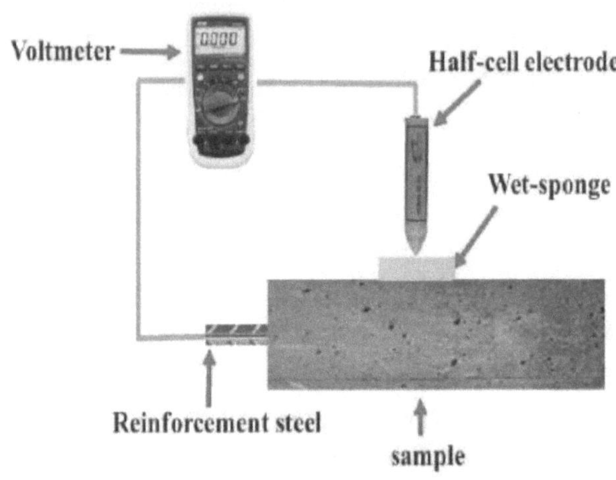


**Figure 26. Schematic of the HCP measurement setup**

### 7.4.2   LPR

The LPR method is a NDT method used to measure the corrosion rate. The data graph obtained from the instrument can be used to calculate the corrosion rate. The polarization resistance measurements are an accurate and rapid technique to measure the rate of corrosion. Typical values relating corrosion measurements to predicted corrosion penetration are given in Table 14 [236].

**Table 14. Values of corrosion measurements**

| Rate of corrosion | Polarization Resistance $R_P(\Omega \text{ cm}^2)$ | Corrosion Current Density $i_{corr}(\mu A/ \text{cm}^2)$ | Corrosion penetration $p$ ($\mu$m/year) |
|---|---|---|---|
| **Very high** | 2.5-0.25 | 10-100 | 100-1000 |
| **High** | 25-2.5 | 1-10 | 10-100 |
| **Low/moderate** | 250-25 | 0.1-1 | 1-10 |
| **Passive** | >250 | <0.1 | <1 |

After the specimens were removed from the chloride water and left to air dry for 24 hours, the LPR test was performed. The specimens were supported on a wooden surface. To conduct the LPR test, Gamry Instruments Reference 600+ potentiostat was used. The cell cable was connected to a reference electrode, counter electrode, rebar and ground. If the surface of the concrete was too dry, samples were pre-wetted. A pre-wetted sponge was used to ensure proper surface contact between the concrete surface and the tip of reference electrode. The equipment was connected to a computer to read the data graph. A complete setup of the Gamry Potentiostat is shown in the Figure 27. The Gamry Echem analyst software was used to run the experiment. This is a single program that runs data-analysis for all type of experiments such as DC corrosions, EIS and physical electrochemistry. Before running the software, the experimental setup values are entered manually (Table 15).

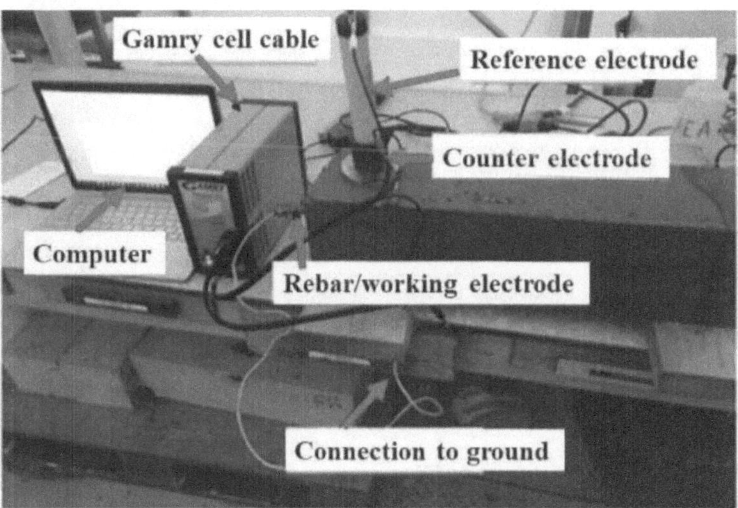


Figure 27. Setup of the Gamry Potentiostat

Same as the HCP technique, a center line with a pre-defined equal spacing of three measuring points at 7 inches distance was marked on the surface of the concrete to obtain a set of reliable test values. The LPR data graphs for these three points were recorded in the computer for both PCC and GPC beams. The values shown in Table 15 were used in the LPR measurement:

**Table 15. Values input in software**

| $E_w$ | Density ($\rho$) | B |
|---|---|---|
| **27.92** | 7.85 g/cm$^2$ | 25 mV |

B value is often taken as 25 mV for active corrosion state and 50 mV for passive conditions. $E_w$ is the equivalent weight of the corroding metal.

The corrosion rate can be computed by using the corrosion current ($I_{CORR}$) generated by the flow of electrons from anode to cathode. By applying the modified version of Faraday's law the following equation can be generated [38]:

$$I_{CORR} = 10^6\, B/R_p \qquad (\mu A/cm^2) \qquad (16)$$

Where $R_p$ is the polarization resistance of a corroding electrode and is defined as the slope of a potential versus current density plot. The dimension of $R_p$ is ohm-cm$^2$.

B is the Stern-Geary coefficient and the Stern-Geary coefficient is given by

$$B = b_a b_c / 2.303(b_a + b_c) \qquad (17)$$

Where $b_a$ and $b_c$ are the anodic and cathodic Tafel slopes.

The corrosion rate in μm per year is given by:

$$CR = 3.27 \times I_{CORR} \times E_w / \rho \qquad (18)$$

Where:

$I_{CORR}$= Corrosion current density in $\mu A/cm^2$

$E_w$ = Equivalent weight of the corroding metal

$\rho$ = density of the corroding metal in g/cm$^3$

## 7.5   Results and discussion
### 7.5.1   Compressive strength

As aforementioned, this section builds on the previous work on the effect of curing regime on compressive strength of GPC performed by authors of the current [36,82,83]. To maintain brevity, this previous work is briefly described here. Authors used steam curing and dry curing methods to accelerate curing of 4″×8″(100×200 mm) cylindrical GPC specimens at five different temperatures including ambient, 30°C, 45°C, 60°C, 80°C. According to the previous results, highest compressive strength (35 MPa at an age of 28 days) was obtained at a temperature of 80°C of steam curing for abovementioned mix proportion, possibly due to the full and uniform internal curing of steam cured specimens. Hence, in the present study, the cylindrical GPC and PCC specimens were steam cured and were tested at 7 and 28 days. A minimum of three GPC and three PCC specimens were used in this test. The average compressive strength of GPC specimens at 7 and 28 days were 26.65 MPa and 31.70 MPa respectively.

For the PCC, the average strength of three samples were 26.93 MPa and 33.67 MPa at 7 and 28 days, respectively. As the age was increased from 7 to 28 days, the compressive strength of GPC and PCC samples increased 18.95% and 25.03% times respectively. It can be concluded from the results that GPC steam cured at 80 °C for 24 hours developed almost an equal strength as the PCC samples cured for 28 days. More importantly, these two mixes were considered comparable for the rest of the research conducted in this work. The mix proportion selected after compression testing of GPC cylindrical samples was used for production of GPC beams. Figure 28 show compressive strength of GPC and PCC at age of 7 and 28 days.

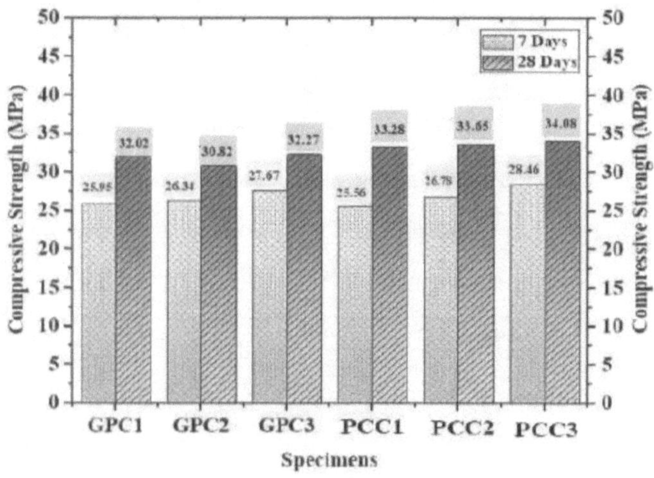


Figure 28. Compressive strength of GPC and PCC

## 7.5.2   NDT evaluation

### 7.5.2.1   Visual inspection

The accelerated corrosion test was terminated at 300 hours and the beams were removed from the chloride solution tank. Figure 29 and Figure 30 show the corroded GPC and PCC specimens after nearly 200 hours of test. The PCC beams started to show signs of rusting after 60 hours of accelerated corrosion exposure. On the other hand, the GPC beams showed no signs of rust for the same period of time. The brown rust stain seen on PCC beams was the first visual evidence of corrosion in the embedded steel. At this stage, it was also observed that corrosion products were floating on the surface of the chloride solution. After nearly 200 hours, a crack was observed in the PCC beams. From this point onwards, the crack propagated longitudinally parallel to the length of the bar. After 300 hours of corrosion exposure, along with a horizontal crack, a vertical crack was also observed which became wider under flexure as shown in Figure 30. On the other hand, there were no cracks observed in the GPC beams indicating better resistance to corrosion when compared to PCC beams.

**Figure 29. PCC beam after 200 hours of test**

**Figure 30. GPC beam after 200 hours of test**

### 7.5.2.2 HCP analysis

Before performing the accelerated corrosion test, the initial HCP readings on day 1 were taken from the voltmeter for both PCC and GPC beams. There were three measurement points on each specimen, and the potential values for these three points were recorded and then averaged. These readings were taken on alternate days until the test reached 300 hours. Figure 31 represents the average HCP values of GPC and PCC specimens. The corrosion process is divided into three stages of propagation i.e., from first 0-3 days as stage 1, from 4 to 8 days as stage 2 and from 8 to 13 days as stage 3. From Figure 31, it is evident that GPC samples exhibited passivity in stage 1 whereas PCC samples exhibited 50 % corrosion likelihood with its potential values raging from -100 mV to -230 mV. Figure 31 also includes the standard [229] for evaluation of corrosion activity. The stage 2 indicates that while GPC samples indicate minimal likelihood of corrosion activity, the PCC samples still indicate 50 % likelihood of corrosion with its potential values ranging from -250 to -300 mV. Further in Stage 3, GPC samples exhibit 90 % probability of corrosion on day 13 while PCC samples exhibit the same on $7^{th}$ day. This clearly distinguishes the fact that GPC offers better resistance to chloride induced damage than PCC. It can be seen that that the drop in potential from day 0 to 3 is very less as it falls after $4^{th}$ day. This is attributed to the fact that a layer of rust gets developed around the rebar that protects it from corrosion progression for a while. As corrosion products build up, iron ions easily diffuse through the layer of rust and results in sharp drop of corrosion potential and quick progression of corrosion. Although, the HCP measurements cannot be directly related with corrosion rate, but it surely signifies the trend of corrosion progression.

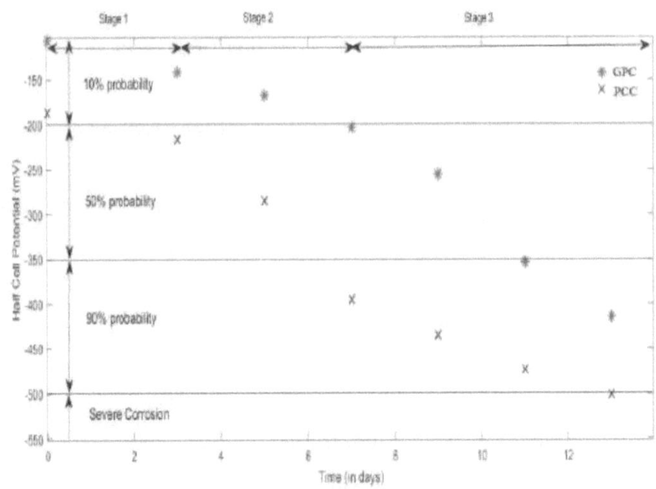


**Figure 31. HCP values of GPC and PCC**

### 7.5.2.3  LPR analysis

The LPR technique is used for accurately measuring the corrosion rates of the specimens and the results are tabulated in Table 16. The corrosion rate is calculated by using the equations discussed in section 2. The experiments are conducted for GPC specimens and PCC specimens using Gamry Potentiostat. The average LPR values of three GPC and three PCC specimens are shown in Figure 32. It can be seen that the recorded LPR of GPC showed higher average values ranging from -380mV to -415mV than PCC (-427mV to-470mV). The average LPR of GPC decreased -8.32% when Im (A) changed from 20 µA to -20 µA whereas average LPR of PCC decreased -9.9% at the same Im (A) range.

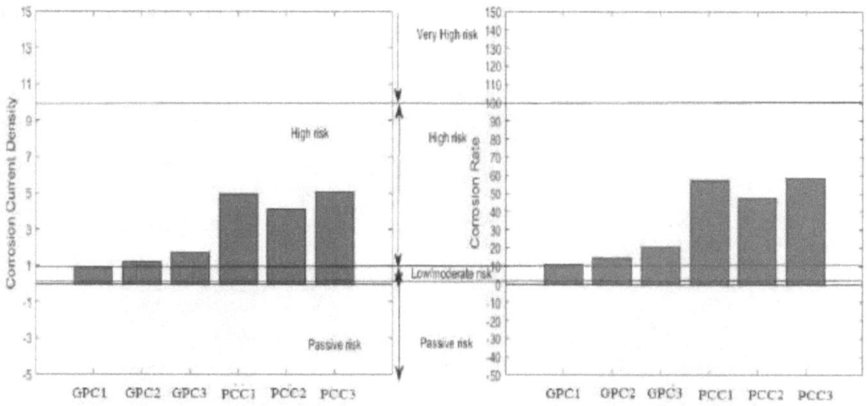



**Figure 32. Average LPR values of GPC and PCC**

**Table 16. LPR resistance test results**

| S.NO. | Specimen | Corrosion Current $I_{CORR}$ ($\mu A/cm^2$) | Corrosion rate ($\mu m/year$) | Corrosion condition |
|---|---|---|---|---|
| 1. | GPC 1 | 0.9113 | 10.598 | Moderate |
| 2. | GPC 2 | 1.2303 | 14.308 | High |
| 3. | GPC 3 | 1.7429 | 20.270 | High |
| 4. | PCC 1 | 4.9214 | 57.233 | Very high |
| 5. | PCC 2 | 4.1010 | 47.696 | Very high |
| 6. | PCC 3 | 5.0471 | 58.698 | Very high |

The corrosion rate of the GPC specimens is in between 10 μm/year and 20 μm/year. This indicates that these specimens have moderate to high rate of corrosion. No cracks were observed on the surface of the GPC specimen, but micro cracks may have occurred in the surrounding areas of the bar due to the corrosion products that was built up inside, which might have allowed some chloride ions to penetrate into the vicinity of the bar. As compared to this, the corrosion rate of the PCC specimens is in between 40 μm/year and 60 μm/year, which indicates very high rate of corrosion.

## 7.5.3 Destructive Tests

### 7.5.3.1 Mass loss measurements

The corrosion assessment of steel bars used in the reinforced concrete can also be done by mass loss measurements [224,237]. The initial mass of each rebar are recorded before the casting procedure. After HCP and LPR experiments, the beams were completely broken to extract the entire rebar as shown in Figure 33.

**Figure 33. Broken PCC and GPC beams**

The rebars were cleaned with deionised water and a metal brush was used to remove the corrosion products from the rebars. The steel rebars visually showed critical corrosion damage for the PCC beams, while the rebars from the GPC beams showed less damage compared to PCC beams. The more negative values observed in GPC beams, however, is not a necessary indication of high risk of corrosion. This can happen due to several reasons such as lack of oxygen at the steel and concrete interface due to oxygen depletion, binding of the chloride ions or lower pH level in GPC compared to PCC binders. The rebars extracted from the beams were weighed and recorded as the final mass. Then the percentage of mass loss was calculated for rebars both in PCC and GPC beams.

**Table 17. Mass loss in rebar in GPC and PCC**

| Specimen | Initial mass (gms) | Final mass (gms) | Mass loss (%) |
|---|---|---|---|
| GPC 1 | 944.5 | 914.9 | 3.13 |
| GPC 2 | 945.5 | 906.4 | 4.13 |
| GPC 3 | 941.9 | 894.5 | 5.16 |
| PCC 1 | 941.9 | 756.8 | 21.80 |
| PCC 2 | 944.3 | 786.3 | 16.73 |
| PCC 3 | 943.7 | 759.4 | 19.52 |

Table 17 shows the percentage mass loss of three GPC and three PCC specimens. The percentage mass loss for the GPC beams was 3.13%, 4.13% and 5.16% respectively, whereas, for the PCC beams it is significantly higher at 21.80%, 16.73% and 19.52% respectively. The PCC beams showed significant mass loss due to the early crack formation which allows the chloride ions to penetrate quickly into the concrete and increase the rate of corrosion.

This finding is in good-agreement with those obtained in previous study [221]. In their studies, the mass loss investigation of steel rebars in GPC and PCC showed that the corrosion rate of steel rebars in GPC is less than PCC, which suggests a higher corrosion resistance of GPC in aggressive environment.

### 7.5.3.2 Residual flexural strength

Corrosion of reinforcement affects the load carrying capacity of any Reinforced Concrete (RC) structure as it reduces the cross-section of rebar that ultimately results in cracking and spalling of concrete. Therefore, it is imperative to study the residual strength of structures after corrosion-induced damage has been taken place. This provides an in depth understanding and insight into the kind of rehabilitation measures that could be utilized for restoring the integrity of the structure. After HCP and LPR testing, the specimens were tested in flexure under 3 point loading using MTI-K testing machine. The residual flexural strength values are given in Table 18. It can be clearly observed from Table 18 that the GPC beam specimen's exhibits improved flexural behaviour than those of PCC specimens. Hence, it can be ascertained that under similar corroding exposure, the load carrying capacity of GPC remains intact for a longer period of time.

**Table 18. Residual flexural strength test values**

| Specimen | Residual Flexural Strength (in MPa) |
|---|---|
| GPC-1 | 5.1 |
| GPC-2 | 4.8 |
| GPC-3 | 4.4 |
| PCC-1 | 3.9 |
| PCC-2 | 3.7 |
| PCC-3 | 3.65 |

### 7.5.4   Discussion

Table 19 gives the average values of 7 and 28 days compressive strength, mass loss, corrosion current and  corrosion rate and average flexural residual strength for all tested GPC as well as PCC samples. There is no significant variation of compressive strength values but there is significant influence of corrosion exposure on of PCC and fly-ash based GPC with additions of bottom-ash. Normally, the compressive strength values of fly-ash based GPC is 5-7 % higher than those of PCC [39], the similar values could be attributed to the addition of bottom-ash and its replacement with flyash itself. The larger particle size of bottom-ash will have some effect on the densification of matrix and consequently, the loss of compressive strength is observed. The various parameters such as pore size, connectivity of pores, shrinkage and movement cracks, chemical characteristics i.e., the chloride binding capacity, alkalinity [237,238] largely controls the dynamics of corrosion of steel in concrete. The durability of a mix majorly depends upon its mineralogical constituents and microstructure [239]; the difference between the reaction products of PCC and GPC could significantly alter their characteristics. Despite the addition of bottom-ash, GPC mix has exhibited significant corrosion resistance than the PCC samples. GPC samples exhibited 78 % less mass-loss, 72 % lower corrosion rate and 73 % lesser corrosion current density. Further, the residual flexural strength is 21 % higher than those of PCC samples. The corrosion resistant behaviour of GPC is probably because of the refinement of pore structures as an outcome of polycondensation reaction and therefore, the filler effect. This further reduces the porosity of GPC mix as the presence of KOH aids in leaching of Si and Al from fly-ash improving the polycondensation and the matrix. The densification of GPC matrix with reduced permeability and improved strength, will restrict the ingress of chloride ions into the steel-concrete ecosystem. This reduces the chloride-diffusion rate in GPC mix as compared to PCC mix which is directly evident from the reduced corrosion current ($I_{corr}$) and corrosion rate as well. The reduced mass loss and less influence of corrosion exposure on GPC could also be attributed to the reduced availability of free chlorides and therefore, increased chloride binding capacity. The slope obtained from the regression of HCP measured values suggest that the rate of drop in potential was much higher for PCC samples than GPC samples. From the Table 19, a clear correlation between all the evaluated corrosion parameters suggest GPC offers enhanced resistant to chloride-induced corrosion.

Table 19. Average test results

| Specimen details | Avg.7day compressive strength | Avg.28day compressive strength | Average Mass loss | Average corrosion rate | Average corrosion current | Average residual flexural strength | Slope of HCP |
|---|---|---|---|---|---|---|---|
| PCC | 26.93 | 33.67 | 19.52 | 54.54 | 4.689 | 3.81 | 28.6 |
| GPC | 26.65 | 34.22 | 4.14 | 15.06 | 1.294 | 4.86 | 20.4 |
| % age Diff | -1.03 | 1.6 | -78 | -72 | -73 | 21 | |

## 7.6 Conclusions

The primary aim of this study was to experimentally study the corrosion resistance of bottom-ash and fly-ash based reinforced GPC, compared to PCC. By analyzing the test results, the following conclusions can be drawn.

- Samples exposed to accelerated corrosion conditions indicate initiated in PCC samples initiated earlier than GPC samples due to a higher resistance of GPC to chloride ingress.

- The HCP values became more negative for both the specimens during the test period. The trend line of both GPC and PCC specimens indicates an increase in the probability of corrosion from day 1 to day 13. After day 13, GPC and PCC both showed 90% probability of corrosion.

- GPC demonstrated lower values for corrosion rate compared to PCC because GPC has moderate to high rate (between 10 µm/year and 20 µm/year). Whereas PCC showed higher rate of corrosion (40 µm/year and 60 µm/year).

- PCC showed higher average mass loss (19.15%) compared to GPC (4.14%). Authors attribute this to existence of higher amount of cracks in PCC. Because cracks cause more chloride ions to penetrate into the paste and increase the rate of corrosion.

- This study has shown a few properties of GPC to enable its use as a building material. GPC possesses a higher resistance to the corrosive activity of salt solutions compared to PCC.

# Chapter 8    Durability of GPC Exposed to Real Environmental Conditions

The following paper has been published in the journal of "**Case Studies in Construction Materials**". It should be noted that this chapter is an adopted and edited version of the following published manuscript.

## 8.1 Introduction

In the past few decades, emission of $CO_2$ through human and natural processes such as human activities, disposal of waste materials and consuming natural resources is considered as the most important cause of climate change. In the construction industry, cement production is one of the main contributors to increasing $CO_2$ emission in the world. That is why the utilization of waste material in the production of concrete can be a possible alternative to cement [240]. According to Mohammadhosseini et al. [241] and Dimitriou et al. [242], greater utilization of waste materials mitigate the negative environmental impacts. Since waste materials are used in numerous applications, it is essential to investigate their effect on the mechanical properties of materials produced. The study by Mohammadhosseini et al. [243] shows that the mechanical properties of Self-Compacting Concrete (SCC) including compressive strength, splitting tensile and flexural strength can be improved by water/binder ratio of 0.4 and 30% replacement of palm oil fuel ash with cement.

Toward achieving a sustainable material and green environment, GPC has been materialized as one of the alternatives to PCC in the construction industry due to its ability to minimize the requirement for natural resources. GPC uses natural aluminosilicate materials activated by an alkali activator, which are then combined with fine and coarse aggregates. Moreover, previous studies [21], [244] on GPC indicate that it can be used for the construction applications due to sufficient durability, workable slump, and comparable grade of strength to its counterpart PCC. The studies by Hardjasaputra et al. [245] and Zerzouri et al. [246] indicate that GPC offers considerable durability and mechanical properties compared to conventional concrete including compressive strength, flexural strength, elastic modulus, sulfuric acid attack resistance etc. Pereira et al. [247] pointed out that both GPC and PCC achieved similar compressive strength (~ 60 MPa) upto 2 years of ageing. In addition, Nabeel et al. [248] reported that GPC synthesized by fly-ash and slag had lower UPV than ordinary Portland cement. However, Nabeel et al. [248] also mentioned that both GPC and PCC with UPV range of 3.5-4.5 Km/s are categorized as "good" quality concrete at age of 7 days and 28 days.

Fly-ash is one of the main constituents of GPC that is pulverized and blown with air into the furnace where it ignites, generates heat and forms ash. Fly-ash is a pozzolan material, containing aluminous and siliceous which forms a compound similar to cement. According to Abualrous et al. [249], the most effective line of attack to increase the waste materials amount in limited lands and reduce $CO_2$ emission is using fly-ash as a supplementary cementitious material or partial substitution for clinker production. It is reported that about 80% of the unburned material or fly-ash is entrained in the flue gas when pulverized coal is burned in a dry condition. The remaining 20% of the ash is dry bottom-ash that has similar chemical properties to fly-ash, consisting of silica and alumina [250]. However, often ignored, bottom-ash as part of the non-combustible residue of combustion in a furnace can also be used to produce a green construction material. That is why, in the current study, attempts have been made to produce GPC made by combination of fly-ash and bottom-ash. Thaarrini and Ramasamy [251] reported that bottom-ash based GPC with NaOH and $Na_2SiO_3$ ratio of 2 yielded a compressive strength of 41.53 MPa and 48.55 MPa under ambient and steam curing (60 °C) conditions respectively.

The utilization of fly-ash or bottom-ash to produce GPC requires an alkaline solution. The most common alkaline solutions used in GPC studies are NaOH and $Na_2SiO_3$ [69,71,75,179,252]. However, a few studies such as Sakkas et al. [183] as well as Hosan et al. [184] reported that the combination of alkaline solution such as KOH and soluble silicate such as $K_2SiO_3$ can also be used in the production of GPC. Sabitha et al. [253] and Bakharev et al. [254] reported that although Sodium-based (Na-based) activators are broadly used for the production of GPC, Potassium-based (K-based) activators are also claimed to decrease the initial setting, improve the geopolymerization process and consequently, increase the compressive strength of GPC as well as its workability. Therefore, in this study attempts have been made to develop a K-based GPC using a unique combination of fly-ash and bottom-ash. Table 20 shows the flow of knowledge on the development of GPC.

Table. 20 The gaps in the knowledge of GPC in real environmental conditions

| References | Gaps/Findings |
|---|---|
| [240-242] | **Findings:**<br>The use of waste materials in production of concrete reduce disposal of waste materials associated with $CO_2$ emission.<br>**Gaps:**<br>Small portion of waste materials was used in production of concrete. However, in the current study, attempts have been made to remove cement from the mixture and use waste materials in production of GPC. |
| [243] | **Gaps:**<br>In this study [243], Palm Oil Fuel Ash (POFA) was used to enhance performance of the mixture. However, the effect of both fly-ash and bottom-ash on durability of GPC is still questioned. That is why in the current study, NDTs were employed to investigate durability of GPC made by both fly-ash and bottom-ash. |
| [21,244] | **Findings:**<br>Both studies showed that GPC is a proper substitute to cement-based concrete which is environmentally friendly construction material. GPC uses low amount of natural resources, emits less $CO_2$ and has superior mechanical properties. |
| [245,246] | **Findings:**<br>The investigation of various properties of GPC including elastic modulus, flexural strength and compressive strength etc. in both normal and aggressive conditions showed that GPC samples are slightly altered compared to its counterpart PCC. |
| [247,248] | **Gaps:**<br>These two studies investigated mechanical properties of GPC made by fly-ash, slag and metakaolin. However, in the current study, attempts have been made to create a durable GPC made by fly-ash and bottom-ash. Moreover, authors could not find any research on mechanical properties of GPC made by fly-ash and bottom-ash in paver block application. |

| References | Gaps/Findings |
|---|---|
| [249,250] | **Findings:**<br><br>Based on the results, fly-ash can be used as a Pozzolan and fine aggregate in concrete. Moreover, this is the most sustainable approach to reducing the amount of fly-ash disposal.<br><br>**Gaps:**<br><br>These studies only compared different types of fly-ash in concrete. However, huge amount of bottom-ash is still disposed all over the world. That is why, the authors of the current study used bottom-ash as one of the main precursors of GPC. |
| [251] | **Findings:**<br><br>This study showed that bottom-ash also can be used to produce a durable GPC with compressive strength range from about 11 to 49 MPa.<br><br>**Gaps:**<br><br>In the current study, the mechanical properties and durability of GPC paver blocks have been investigated in real environmental conditions and compared with its counterpart PCC paver blocks. |
| [71,184,254] | **Gaps:**<br><br>It is well-known that Na-based solution is basically used in production of GPC. However, not much work reported on the mechanical properties of K-based GPC paver blocks in real environmental conditions. In the current study, NDTs such as UPV were used to investigate durability of K-based GPC paver blocks. |
| Contribution of this current study | As aforementioned, the mechanical properties and durability of fly-ash and bottom-ash in GPC have been explored by many researchers [240,254]. However, not much information is available on the durability and leaching of chemical metals of K-based GPC made by combination of 50% fly-ash and 50% bottom-ash in real environmental conditions association with real traffic load. That is why, this study creates a paradigm for future practical execution of this new generation of concrete called GPC. |

## 8.1.1  Research significance

Since the consumption of bottom-ash is comparatively limited, utilizing this by-product material is one of the critical challenges in recent decades [25,255]. Hence, one of the main objectives of the current study is to develop a K-based GPC made by a combination of 50% fly-ash and 50% bottom-ash. Further, the knowledge gap on mechanical properties and leach-ability under real environmental exposure is filled. In this study, as a macro-scale investigation, GPC paver block has been selected as an application since there is no work authors could find that reports the effect of low temperature and wetting-drying on K-based GPC's performance under real environmental conditions. As a laboratory investigation, Khandol et al. [256] pointed out that fly-ash based alkali activated paver block has superior compressive strength, abrasion resistance, and lower water absorption compared to its counterpart PCC.

Leaching assessment of GPC is another important concern and matter of interest in engineering concepts, which must be known when using this material in exposed applications. Numerous studies [257-259] have assessed the leach-ability of PCC, however, there is no information available for K-based GPC made by a combination of fly-ash and bottom-ash. Moreover, tests in real environmental conditions including cold climates to estimate the interaction of K-based GPC leaching with ground-water are not reported. Hence, another main objective of the present study is to investigate the leach-ability of chemical metals of both PCC and GPC over every thirty days of exposure. Izquierdo et al. [260] reported that fly-ash based geopolymer matrix is appropriate for the immobilization of many chemical elements including Be, Bi, Cd, Co, Cr, Cu, Nb, Ni, Pb, Sn, Th, U, Y, Zr. Though, the leaching amount of some elements such as B, Mo, Se, V were higher in the geopolymer matrix as compared with fly-ash (raw material). Several knowledge gaps that were indicated in this section are studied in this current work.

## 8.2 Precursors of GPC

### 8.2.1 Fly-ash and bottom-ash

The properties of procurers have been already explained in section 4.5.1. Table 21 shows the comparison between chemical compositions of Lafarge fly-ash and other types of fly-ash from other sources. As can be seen in Table 21, all the available chemical elements of three types of fly-ash are in the range of ASTM C618 [97]. Comparing the chemical contents of three types of fly-ash, CaO was found in a greater percentage in Lafarge fly-ash which due to the polymerization reaction with hydrated materials results in a greater compressive strength [261].

**Table 21. Chemical compositions of Lafarge fly-ash and other types of fly-ash**

| Properties | Fly-ash (%) Current study (Lafarge) | Fly-ash (%) [262] | Fly-ash (%) [263] | ASTM C618 (%) [97] |
|---|---|---|---|---|
| $SiO_2$ | 47.1 | 62.04 | 59.7 | |
| $Al_2O_3$ | 17.4 | 25.5 | 27.51 | 70 (min) |
| $Fe_2O_3$ | 5.7 | 4.28 | 4.91 | |
| CaO | 14 | 3.96 | 1.45 | N/A |
| MgO | 5.4 | 1.27 | 1.18 | N/A |
| $SO_3$ | 0.8 | 0.73 | 0.16 | 5.0 (max) |
| LOI | 0.19 | N/A | 2.66 | 6.0 (max) |

| Properties | Fly-ash (%) Current study (Lafarge) | Fly-ash (%) [262] | Fly-ash (%) [263] | ASTM C618 (%) [97] |
|---|---|---|---|---|
| $Na_2O$ | N/A | 0.46 | 0.82 | N/A |
| $K_2O$ | N/A | N/A | 2.39 | N/A |
| $TiO_2$ | N/A | 1.33 | N/A | N/A |
| $P_2O_5$ | N/A | 0.31 | N/A | N/A |
| $Mn_2O_3$ | N/A | N/A | N/A | N/A |

The physical properties of fly-ash already shown in Table 2 were analyzed at the Lafarge Seattle Concrete Lab. It can be seen in Table 2 that all the physical properties of used fly-ash in this study complies with the requirement of ASTM C618 [97].

Bottom-ash with coarser particle size compare to fly-ash was sieved (#1.18mm) to remove large particles to increase the surface area of particle used to ultimately help to achieve reasonable strength. Table 22 shows the value of various chemical elements of three types of bottom-ash.

According to Table 21 and Table 22, the chemical compositions of both fly-ash and bottom-ash obtained from various sources are varied. It can be attributed to the chemical content of the coal burned and the type of operation process [264]. It should be noted that there is no standard reported the specification/limitation for utilization of bottom-ash in a concrete mixture.

Table 22. Chemical compositions of Lafarge bottom-ash and other types of bottom-ash

| Properties | Bottom-ash (%) Current study (Lafarge) | Bottom-ash (%) [264] | Bottom-ash (%) [265] |
|---|---|---|---|
| $SiO_2$ | 60.11 | 22.90 | 19.12 |
| $Al_2O_3$ | 14.35 | 0.18 | 12.03 |
| $Fe_2O_3$ | 5.92 | 11.39 | 9.31 |
| $CaO$ | 10.40 | 17.93 | 43.11 |
| $MgO$ | 4.49 | 1.0.4 | 2.11 |
| $SO_3$ | 0.10 | 0.73 | 2.39 |
| $LOI$ | 0.00 | N/A | N/A |
| $Na_2O$ | 2.232 | 3.68 | 2.35 |
| $K_2O$ | 1.766 | N/A | 0.84 |
| $TiO_2$ | 0.892 | N/A | 2.48 |
| $P_2O_5$ | 0.200 | N/A | 2.62 |
| $Mn_2O_3$ | 0.093 | N/A | N/A |

The range of other properties of both fly-ash and bottom-ash obtained from the Material Safety Data Sheets (MSDS) of Lafarge Canada Inc have been already shown in Table 3. The physical properties of bottom-ash such as soundness are not precisely measured by Lafarge Canada Inc. because bottom-ash is not typically used in the construction industry.

Traffic is one of the most factors in paver block design. The deterioration caused to paver blocks by traffic depends on weight of the vehicles and number of load repetitions over the traffic analysis period. In order to quantify this deterioration, the number of Equivalent Single Axle Loads (ESAL) should be calculated. IS 15658 standard [266] recommended different grade of paver blocks for various construction areas and traffic categories. In accordance to IS 15658 standard [266], the grade of M35 ($f'_c$=35 MPa) was considered for this study because less than 150 commercial vehicles daily used the parking area. The concentration of 12 Molarity (M) and 50:50% mass ratio of bottom-ash to fly-ash were selected to achieve target strength of 35 MPa. This target strength was chosen to produce GPC paver blocks with properties comparable to that of commercially available PCC paver blocks and to later compare the durability of these two types of paver blocks in real environmental conditions. However, the proprietary mix design of PCC paver blocks was not revealed by the manufacturer. Commercially produced PCC paver blocks were purchased from Abbotsford Concrete Products Manufacturer of Abbotsford, B.C., Canada.

### 8.2.2 Morphology of fly-ash and bottom-ash

A SEM was performed using Hitachi S-4800 to capture images of fly-ash and bottom-ash at the Advanced Microscopy Facility (AMF) of the University of Victoria. A thin layer of carbon was sputtered on the surface of ashes to increase the conductivity of particles. The SEM accelerating voltage of 15.0KV was used.

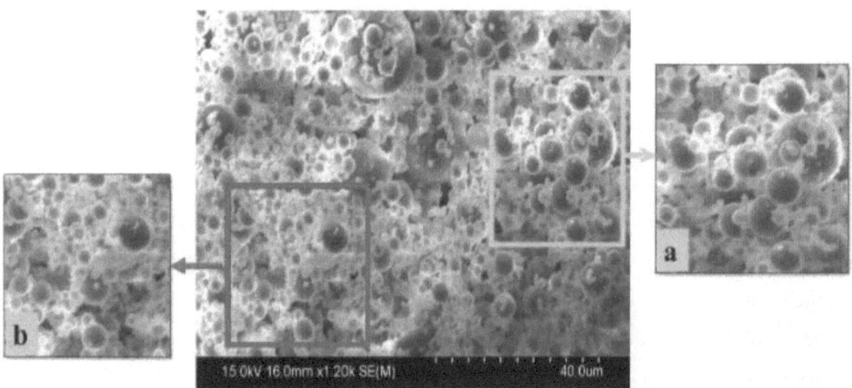


**Figure 34. SEM micrograph of fly-ash particles**
**(a) & (b) Two random areas selected to characterize the shape of fly-ash particles**

Basically, fly-ash particles are divided into three classes: solid sphere, cenospheres and plerospheres. Cenospheres fly-ash has lower bulk density, greater thermal resistance, higher workability and superior strength [267]. While, plerospheres are created by impregnated cenospheres with micro-spheres, which is why plerospheres are heavier than cenospheres due to extra weight of bagged microspheres [268].

As can be seen in Figure 34 (a) & (b), fly-ash particles are glassy, spherical in shape and there is no evidence of cenospheres (totally hollow) and plerospheres (filled with several tiny spheres) shapes. Generally, lower fineness and low carbon content in fly-ash decrease the water demand of concrete [269] due to its specific surface role [270]. Table 2 shows that Loss on Ignition (LOI) of used fly-ash in the current study is 0.19%. Fly-ash with less than 4% LOI is considered as low carbon content fly-ash. Figure 34 (a) & (b) present that all fly-ash particles are spherical glassy. The spherical glassy shape of fly-ash particles is also reported by Davidovits [3].

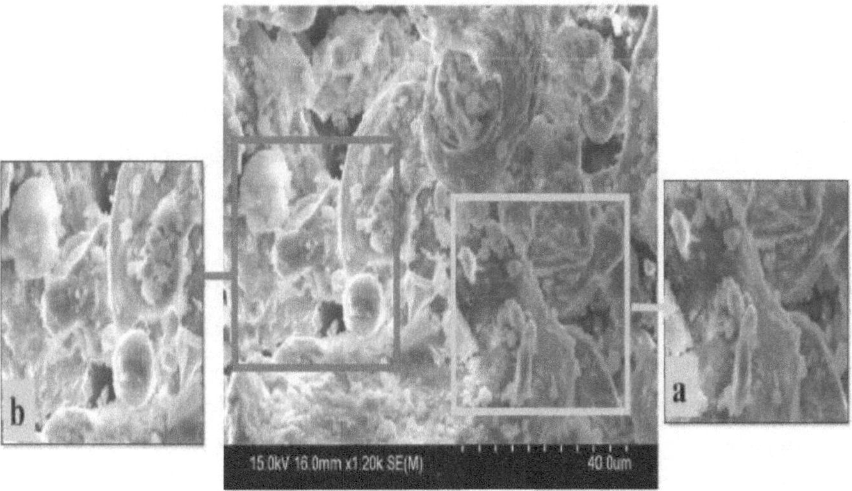


**Figure 35. SEM micrograph of bottom-ash particles**
**(a) & (b) Two random areas selected to characterize the shape of bottom-ash particles**

A magnification of 1200x is used both in Figure 34 and Figure 35 for easy comparison. Based on these observations, it is expected that replacement of fly-ash with any amount of bottom-ash would negatively impact on workability, since not all bottom-ash particles were found to be spherical.

Figure 35(a) and Figure 35(b) were randomly selected to characterize the shape of bottom-ash particles. It can be seen that at the accelerating voltage of 15.0 KV, non-uniform shapes with a small amount of spherical shape were detected in bottom-ash samples.

In accordance to a previous study on microstructure of fly-ash and bottom-ash performed by the current authors [98], the particle size of fly-ash was about 1–20 µm with average size of 10 µm. While, the mean size of the rounded shape bottom-ash particles was 58.53 µm. The average height and width of the irregular shaped bottom-ash particles was 22.59 µm and 12.78 µm respectively.

### 8.2.3   Alkaline solution

The properties of alkaline solution have been already explained in section 4.5.2.

### 8.2.4   Aggregates

The particle size distribution of coarse aggregates and fine aggregates has been already explained in section 4.5.2

### 8.3   Method of casting and curing

### 8.3.1   Specimen preparation

In this study, 300 GPC paver blocks were cast at the Civil Engineering Materials Facility at the University of Victoria in accordance with ASTM C192 [103]. The production of GPC has been already explained in section 4.6.1.

### 8.3.2   Curing methods

Various trial mix designs were created to achieve a strength target of 35 MPa. Different curing temperatures and durations tested on cylindrical samples were selected to develop GPC since the curing temperature and duration is a critical parameter in the geopolymerization process to obtain higher strength. Two types of accelerated curing methods were used in this study:

#### 8.3.2.1   Dry curing

The dry curing method has been already explained in section 4.6.2.

#### 8.3.2.2   Steam curing

The steam curing method has been already explained in section 4.6.2.

**Figure 36. Raw materials, curing temperature, shape and size of GPC**

Figure 36(a) shows all the raw materials used in the production of GPC. Figure 36(b) also shows the immersed paver block into water and cured at 80 °C to achieve higher compressive strength. Figure 36(c) shows the cured paver block and Figure 36(d) shows the size, color and shape of GPC after 28 days of curing.

## 8.4 Effect of geometry

Since the geometry of specimens affects the ultimate compressive strength of concrete [271, 272], the compressive strength of cylindrical-shaped samples was compared to the compressive strength of rectangular-shaped samples (paver blocks) in accordance with British Standard EN-206 [273].

According to this standard, compressive strength test is performed on either 150×300mm cylindrical or 150 mm cubical samples. As an example, when the minimum specified compressive strength of a cylindrical-shaped sample for lightweight concrete is 35MPa, the cubic sample is expected to be 38 MPa [273]. However, it is mentioned in British Standard EN-206 [273] that compressive strength for other sizes of samples should be estimated for non-representative values. Based on the above-mentioned discussion, in this project, the 100×200mm cylinder strength was correlated to the paver block strength by performing laboratory tests. These results are presented later in this paper.

## 8.5 Field placement details

Two areas at the entrance of parking lot#3 at the UVic were selected to place a total of 150 GPC and 210 PCC paver blocks. Area 1 shown in Figure 37(a) was designated to investigate the combined effect of traffic load and low temperature on properties of GPC and PCC. Figure 37(b) (Area 2) indicates the designated area for studying the effect of low temperature only on the RDME and leach-ability of GPC and PCC. Paver blocks in this section were not expected to be exposed to any traffic load.

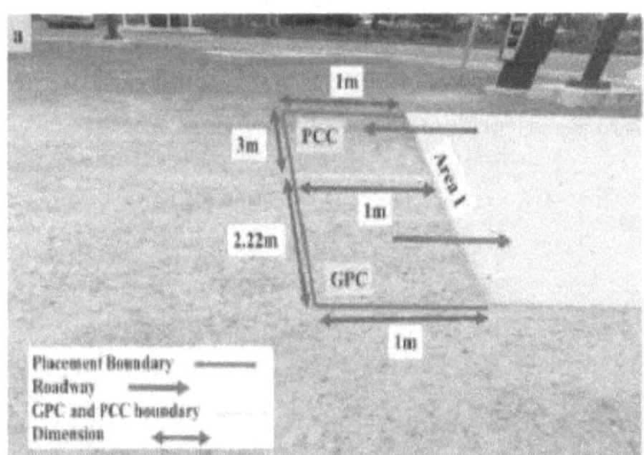

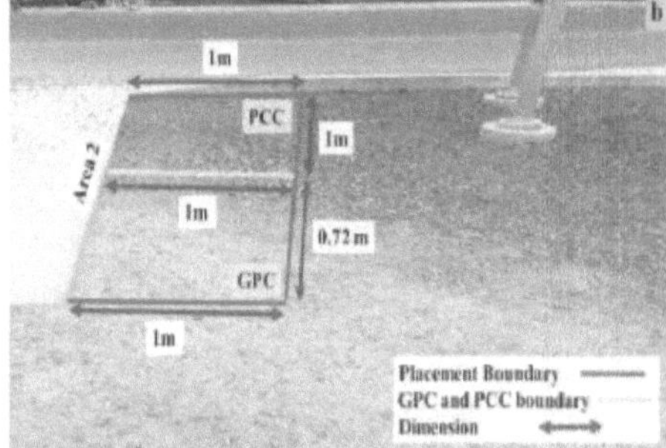

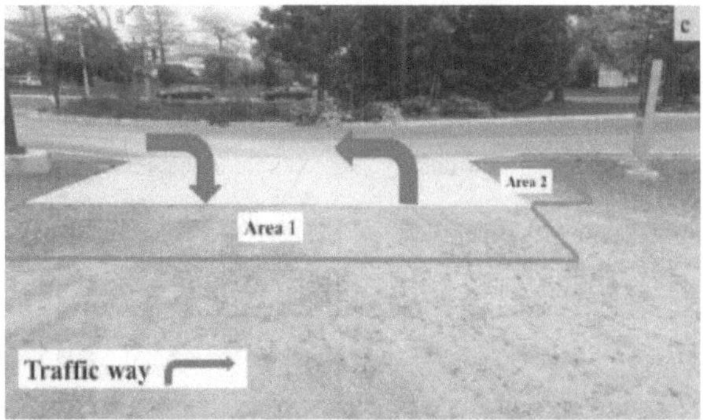


**Figure 37. Site details**
**(a)  Area for effect of low temperature & traffic load investigation**
**(b)  Area for low temperature, RDME & leach-ability**
**(c)  Traffic way**

Figure 37(a) shows the placement area of $2.22 \times 1$ m$^2$ and $3 \times 1$ m$^2$ for GPC and PCC paver blocks respectively. Figure 37(b) also indicates that the GPC and PCC placement size is $1 \times 0.72$m$^2$ and $1 \times 1$m$^2$ respectively. Traffic shown in Figure 37(c) was expected to cross straight over the paver blocks and to turn over an existing concrete slab. So, both paver types were exposed to more or less the same conditions.

## 8.6   Test methods

Initial testing on GPC cylindrical samples was performed during the mix optimization phase. As mentioned earlier, the highest compressive strength was achieved at a temperature of 80 °C of steam curing. So, samples for all subsequent tests were cured at a temperature of 80 °C. In this research, NDT devices such as rebound hammer and UPV were also employed to evaluate the properties of GPC and PCC exposed to real environmental conditions.

### 8.6.1   Compressive strength of cylinders

A total of 54 GPC cylindrical samples (100 mm diameter and 200 mm height) were tested in accordance with ASTM C39 [105] using the Forney compression testing machine #AD 650. According to ASTM C39 [105], the compressive load on the specimen was applied at a rate of $0.25 \pm 0.05$ MPa/s.

### 8.6.2 Compressive strength of paver blocks

According to ASTM C805 [274], Schmidt hammer or rebound hammer processes the rebound of a spring-loaded mass impacting against the surface of a concrete sample. The Schmidt hammer knocks the concrete surface at certain energy. Its rebound is dependent on the rigidity of the concrete surface/sample. The rebound rate can be used to measure the concrete's compressive strength by using a conversion chart. In this study, Original Proceq Schmidt hammer Type N/NR, as a surface hardness method, was used to measure the change in compressive strength of paver blocks every month after initial placement in accordance with ASTM C805 [274]. The tests can be carried out in three positions including horizontal, vertically (upward or downward) and angled position. In this study, Schmidt hammer was hit on the top surface of the paver blocks vertically downward to measure their in-place compressive strength. According to ASTM C805 [274], an average of ten Schmidt hammer readings was calculated to obtain accurate results. Moreover, Schmidt hammer was calibrated to be functioning properly after each thirty reading using an ideal test anvil.

### 8.6.3 UPV

UPV is one of the effective methods to determine uniformity and quality of concrete. The UPV method evaluates the travel time of longitudinal ultrasonic waves passing through the concrete. The path length between transducers divided by the travel time gives the average velocity of wave propagation. UPV is used to check the existence of internal flaws and voids in accordance with ASTM C597 [275]. Velocity reduction of UPV test results shows the internal deterioration of the specimens [276]. In this study, the Pundit Lab Proceq UPV test instrument was used with bandwidth, measuring resolution, pulse voltage UPV and nominal transducer frequency of 20-500 kHz, 0.1 us, 125-500 V and 24-500 kHz respectively. Indirect transmission method (transducers were held on the same surface) was used to evaluate the velocity of paver blocks in-situ. Since the paver blocks were placed on bedding sand and since UPV of sand is different from that of UPV of paver blocks, the optimum pulse width, length and height parameters in the device were set to the dimensions of paver blocks so as to only measure the UPV of paver blocks and eliminate any effect of its surroundings. Every month, three readings were recorded from each paver block to have accurate data of their UPV.

The following equation was used to predict the dynamic modulus of elasticity form measured UPV in accordance with ASTM C597 [275]:

$$E = \frac{\rho\,(1+\mu)(1-2\mu)V^2}{1-\mu} \qquad (19)$$

E= dynamic elastic modulus, GPa

$\rho$ = density in kg/m$^3$ (Average measured density of GPC: $\rho$=2439 kg/m$^3$ and average assumed density of PCC:  $\rho$= 2400 kg/ m$^3$)

V= measured pulse velocity in m/second

$\mu$= assumed dynamic Poisson's ratio ($\mu$=0.24 for both types of concrete)

### 8.6.3.1  Seasonal humidity effects

The humidity of materials changes their response to ultrasonic waves passing through them. Hence, it is important to investigate the effect of seasonal humidity on the velocity of paver blocks since the paver blocks are exposed to wetting-drying cycles and different humidity conditions. Generally, in the field, all the paver blocks experienced three humidity conditions: fully wet, Saturated Surface Dry (SSD) and dry. So, the same laboratory conditions were simulated to measure the effect of humidity on velocity and compressive strength of paver blocks. These results are reported later.  Figure 38 shows the average of high and low temperatures and average humidity recorded in the city of Victoria from November 2017 to August 2018 [277]. The Schmidt hammer and UPV test were performed over 240 days with the lowest and highest temperatures of 1°C to 20 °C respectively. These results are also reported later.

The environmental temperature conditions include mainly the freezing temperature (range) and the freezing rate. The lowest temperature determines the attainable pore size of freezing, and thus the quantity of ice formation. Larger freezing ranges involve more pores in freezing. Note that natural frost action has a freezing temperature rarely below −20 °C. The freezing rate is believed to play a role in the frost damage and faster freezing is supposed to cause more damage than slower freezing. Based on the hydraulic pressure theory, the magnitude of the hydraulic pressure in pores is proportional to the freezing rate [278]. The results in the literature seem to confirm, if not the proportionality between frost damage and freezing rate, larger spacing factors of air voids could be tolerated at lower freezing rates. The freezing rate involves certainly the kinetics aspect of pore freezing and the viscous flow of liquid water as well. Note that the natural freezing rate is rarely above 2 °C/h, but the rate in laboratory accelerated tests can reach 20 °C/h [278]. A higher freezing rate, 12 °C/h, induces a larger freezing expansion, but the most severe damage, in terms of residual strain after each freeze-thaw cycle, occurs at the lowest freezing rate, 2.5 °C/h.

Under actual exposure condition, the temperature changes are more gradual and in cold areas the concrete may be frozen for many days. Failure of concrete in service, especially if frozen for several weeks, is likely to involve ice accretion. However, the relative performance of concretes and aggregates in the test appear to adequately predict their relative performance in the field, which is of great assistance in selecting durable concrete. Some concretes have been found to only have fair performance in the test but still have adequate in-service frost resistance. If concrete has a high durability factor after 300 cycles (defined cycle by ASTM C666 [9]), it should be able to withstand a severe freezing and thawing environment [9]. The highest temperature ever recorded at Victoria was 36.0 °C (96.8 °F) on 11 July 2007, however the University of Victoria reached 37.6 °C (99.7 °F) on 29 July 2009 [277]. The coldest temperature on record is −18.9 °C (−2 °F) on 31 January 1893 [277]. As aforementioned, the temperature in Victoria rarely goes below the defined temperature (-18 °C) by ASTM C666 [9].

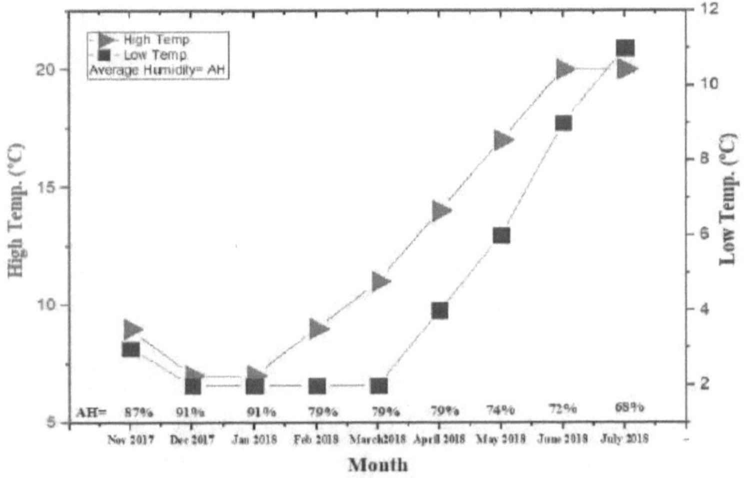


**Figure 38. Highest and lowest temperature over months** *[277]*

### 8.6.4 RFT

RFT or RTG as a new NDT method has brought a lot of attention to determine material deterioration. However, this method is typically only performed in a lab environment. While, UPV method can be used either on-site or in the laboratory. So, a comparison between RDME based on RFT and UPV was made to develop a better understanding of any correlation that exists between these two methods. RFT/RTG method is based on the determination of fundamental transverse, longitudinal and torsional resonant frequencies of vibration of a concrete sample that is created by an impact and sensed by an accelerometer.

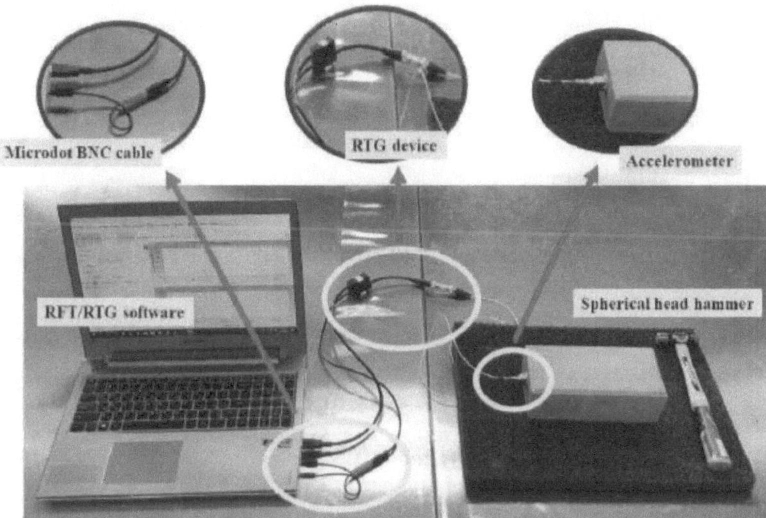


**Figure 39. The main components of RTG/RFT equipment**

Figure 39 shows different parts of the RTG/RFT device which is designed for laboratory usage. Transverse impact RFT and UPV test were performed on thirty paver blocks (in the lab) to measure the dynamic modulus of elasticity. In this study, Olson RFT/RTG instrument consists of a USB powered RTG device, BNC connection, 2 oz. ball-peen hammer (spherical head hammer), an accelerometer (10 mV/g) and Windows 7-10 device running Olson instruments' RFT/RTG software were used. The following equation was used to predict the dynamic modulus of elasticity in accordance with ASTM C215 [39]:

$$G_d = CMn^2 \qquad (20)$$

Where

$G_d$= dynamic elastic modulus, GPa,

$C$= 0.9464 $(L^3T/bt^3)$, $m^{-1}$,

$L$= length of GPC prism, m,

$M$= mass of GPC cylinder (kg),

$N$= fundamental transverse frequency, Hz,

$t$, $b$ = dimensions of cross-section of a prism, m,

$T$= correction factor

### 8.6.5  Water quality assessment

Chemical activators and waste materials contain heavy metals that can be mobilized and can leach into the environment [279]. Hence, the leach-ability of GPC and PCC paver blocks were measured to trace metals in accordance with the United States Environmental Protection Agency Standard 1311 [280]. The leaching tests were performed using HACH strips to determine the quality of collected sample water from the site.

### 8.7  Results and discussions
### 8.7.1  Laboratory investigation of GPC and PCC samples
### 8.7.1.1  Compressive strength of cylindrical GPC samples

Dissolution and geopolymerization of alumina-silicate gel of GPC significantly depend on the curing condition. Both higher curing temperature and duration can be used to accelerate the polycondensation process [42,281]. Steam curing and dry curing methods were used to accelerate curing of GPC at five different temperatures including ambient, 30°C, 45°C, 60°C, 80°C. The compressive strength of dry-cured and steam-cured GPC specimens was already shown in Figure 4. The Figure 4 showed that highest compressive strength was achieved at a temperature of 80°C of steam curing. It was also observed that the compressive strength of steam-cured samples and dry-cured samples increased by about 3.5 and 2.3 times respectively when the curing temperature was changed from ambient temperature (~10°C) to 80°C. Figure 4 also shows that for the type of GPC developed in this study, a minimum increase in strength is realized when curing temperature is increased from ambient to 30°C. Compressive strength almost doubles when GPC is steam-cured at 45°C as opposed to 10°C. At all temperatures, steam curing outperforms dry curing. The authors attribute this to full and uniform internal curing of steam-cured samples. This study also indicates that reasonable strength can be achieved at 60°C of curing and for some applications it may not be necessary to cure at high temperatures such as 80°C. However, the target compressive strength of 35 MPa for the current application/construction area (parking lot with light traffic) was achieved at 80°C after initial trial experiments on GPC samples. The abovementioned finding is in good-agreement with a performed study by Noushini et al. [89] that Na-based GPC made only with fly-ash achieved a higher rate of compressive strength ranged from 27.4-62.3 MPa when samples were exposed to elevated temperature (23-90 °C). However, it should be noted that Na-based GPC made only with fly-ash slightly reached to higher compressive strength than K-based GPC made by fly-ash and bottom-ash (present study) cured to the same temperature and duration. It can be attributed to lower activation potential of KOH compared to NaOH which is because of the ionic diameter difference between sodium and potassium [282].

### 8.7.1.2 Compressive strength of both GPC and PCC paver blocks

Six paver blocks were cast and later cured at a temperature of 80°C to find the compressive strength ratio between rectangular-shaped and cylindrical-shaped specimens. The samples were placed horizontally between two plates of the compressive testing machine. As can be seen in Figure 40, the GPC paver block samples give an average compressive strength slightly lower than the cylindrical samples with the same curing method possibly due to higher stress concentration at corners of paver blocks [283]. This result shows that British Standard EN-206 [273] overestimated the compressive strength of the GPC paver block. One reason for this discrepancy can be attributed to the fact that this standard was originally designed for calculating the compressive strength of PCC and not GPC.

The average compressive strength and geometry conversion factor of GPC paver blocks were about 31.4 MPa and 0.89 respectively. The average compressive strength of PCC paver blocks was 33.5 MPa. No conversion factor for PCC paver blocks could be calculated as commercially available PCC paver blocks were used and no cylinders could be cast. However, it is reported by Ngoc et al. [284] that the mean compressive strength conversion factor of cubic samples (size 150 mm) and cylindrical compressive strength (150x300 mm) is 0.84.

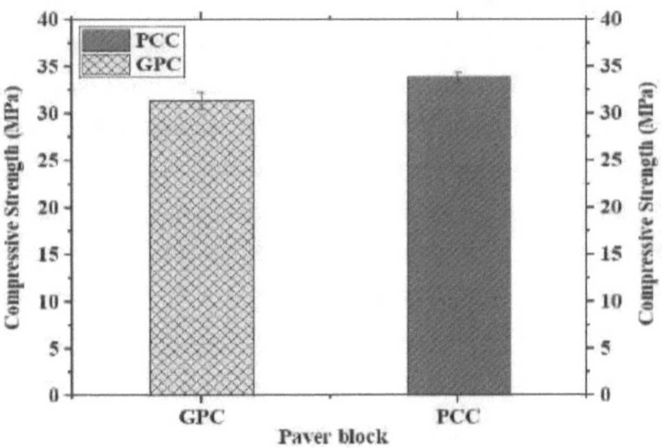

Figure 40. Average compressive strength of GPC and PCC paver block

### 8.7.1.3 Effect of humidity on mass and velocity of GPC and PCC

According to ASTM C597 [275], the humidity level of concrete samples affect the pulse velocity. Laboratory simulation of seasonal conditions was performed to measure the mass and velocity of paver blocks at different humidity levels to have a better understanding of the effect of humidity and wetting-drying cycles on the velocity of paver blocks in real environmental conditions. Six GPC and PCC paver blocks were immersed into water to measure the mass and UPV of SSD and fully wet samples (soaked for 14 days).

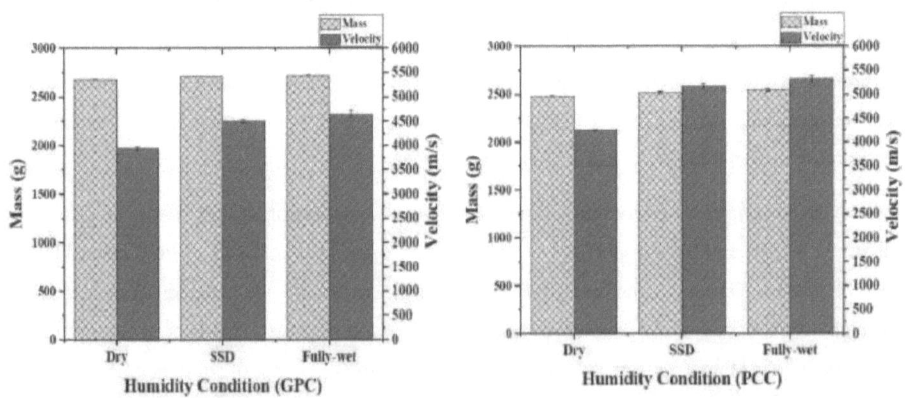

Figure 41. Effect of humidity on mass and velocity of GPC and PCC

Figure 41 shows the average mass and velocity of GPC and PCC paver blocks in three humidity conditions. The GPC samples showed that the mass and velocity ratio between dry and fully wet conditions is about 0.98 and 0.84 respectively. While, it can be observed from the results of PCC samples that the mass and velocity ratio between dry and fully wet conditions is about 0.97 and 0.79 respectively. So, for both types of concrete, the velocity and mass increased with a higher humidity level as would be expected. It is also noted that the susceptibility for mass and velocity to increase with an increase in moisture was similar for both types of concrete.

### 8.7.2    On-site compressive strength and velocity

Figure 42 is a schematic showing the placement of PCC and GPC paver blocks on-site. In area 1, codes 1-8 are for GPC and codes 9-22 are for PCC paver blocks. In area 2, codes 1-3 are for GPC and codes 4-8 are for PCC paver blocks. The codes/numbers are later used as labels in Figure 43 and Figure 44. The total number of paver block rows for GPC and PCC are shown with red color and blue color respectively. Three rows shown with green hatch were considered just for investigation of the effect of traffic load, but it is beyond the scope of this thesis.

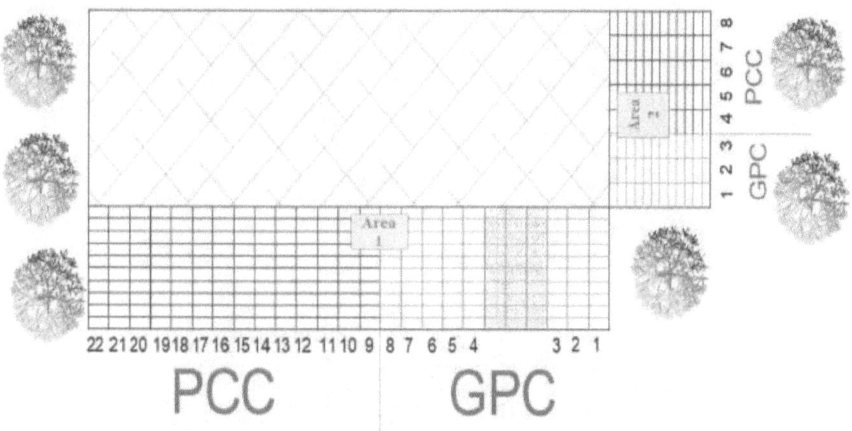



Figure 42. Top view of the site and coding pattern

The relationship between compressive strength, average velocities of each GPC and PCC paver blocks are shown in Figure 43 and Figure 44. The compressive strength and velocity of paver blocks were measured every month after the initial placement.

### 8.7.2.1  Compressive strength and velocity of GPC

From visual inspection, over the first 60 days, small local deterioration of surface (scaling) of some GPC paver blocks was observed. This action led to large cracks extending downward from the damaged zone to the bottom of GPC paver blocks (15 blocks). Over the following 60 days, more voids were available for organic and inorganic solutions to permeate through the GPC paver blocks and produce more spalling in GPC paver blocks. After 120 days of exposure to real environmental conditions, the compressive strength and velocity decreased and finally, after approximately 240 days, 15% of GPC paver blocks showed reduced strength and velocity and consequently, their testing was terminated. Of the total placed, 85% of paver blocks indicated acceptable results.

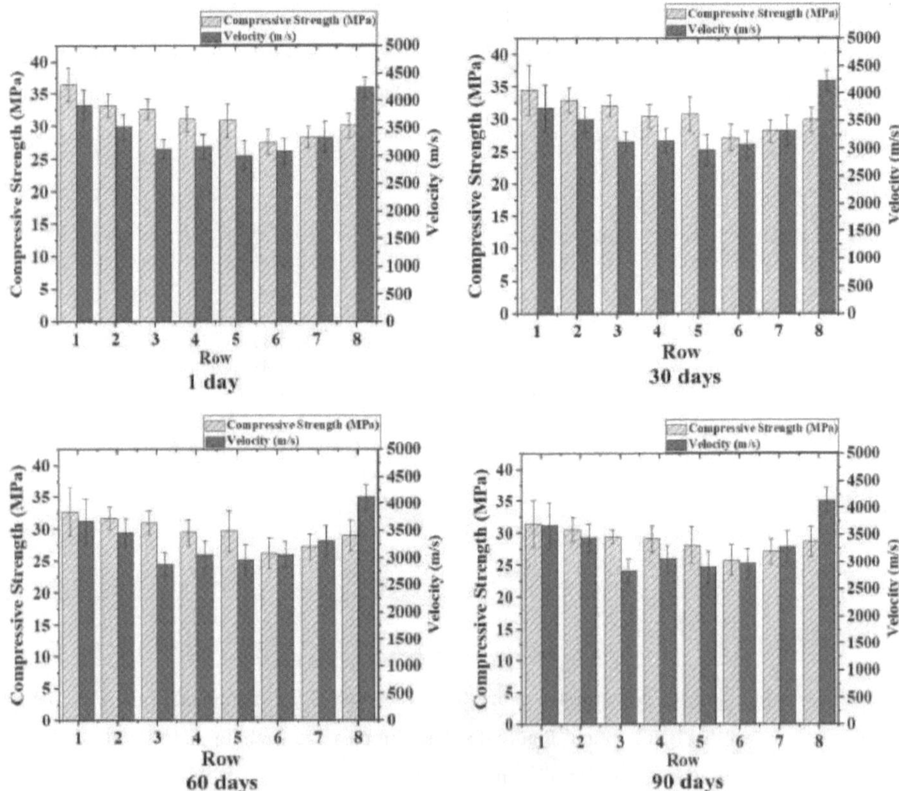

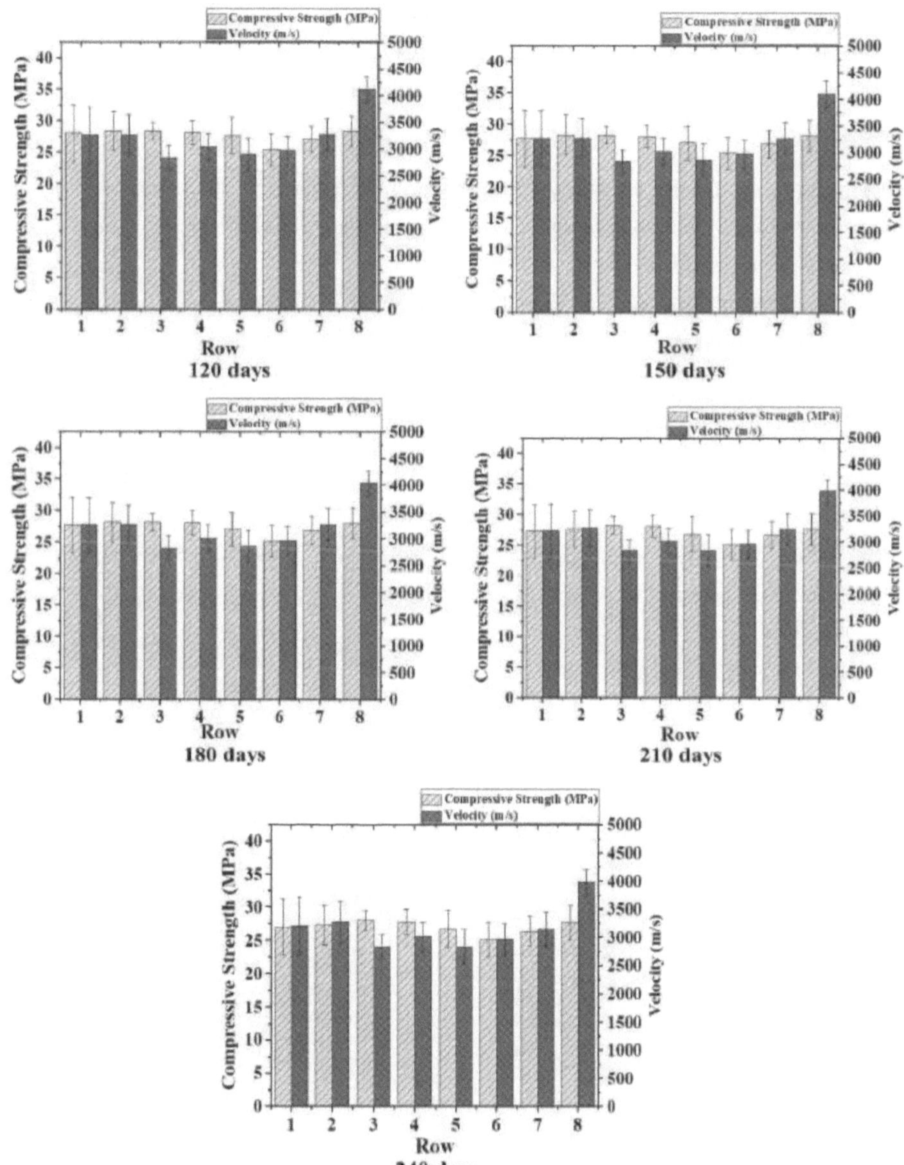

**Figure 43. Average compressive strength and velocity of paver blocks over 240 days of exposure (1-8 rows)**

In this study, attempts were made to take all readings of UPV and compressive strength on a single day to minimize the effect of changing environmental factors. Figure 43 shows average compressive strength and velocity of 35 GPC and 59 PCC paver blocks and Figure 44 shows average compressive strength and velocity of 80 GPC and 140 PCC paver blocks.

Both Figure 43 and Figure 44 show the change in average compressive strength and velocity of paver blocks over 240 days of exposure. As can be seen in Figure 43 and Figure 44, on an average, the compressive strength and velocity of GPC paver blocks were higher than PCC paver blocks on the first day of experiments compared to 150 days of exposure; compressive strength and velocity of GPC decreased progressively over 150 days of exposure. However, it can also be seen that the compressive strength and velocity of both types of paver blocks decreased slightly from 150 days to 240 days of exposure.

Figure 43 shows that when the age of exposure increased from 1 to 240 days, the average velocity of GPC paver blocks (1-3) decreased from 3523.2 to 3099.3 GPa whereas the average velocity of PCC (4-8) paver blocks decreased from 3370.8 to 3192.8 GPa. It also can be seen in Figure 43 that the average compressive strength of GPC (1-3) and PCC (4-8) decreased by 19.51% and 9.65% respectively.

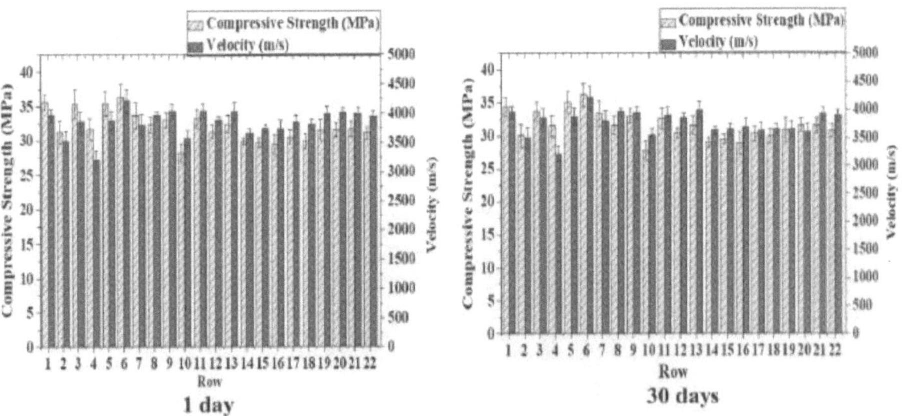

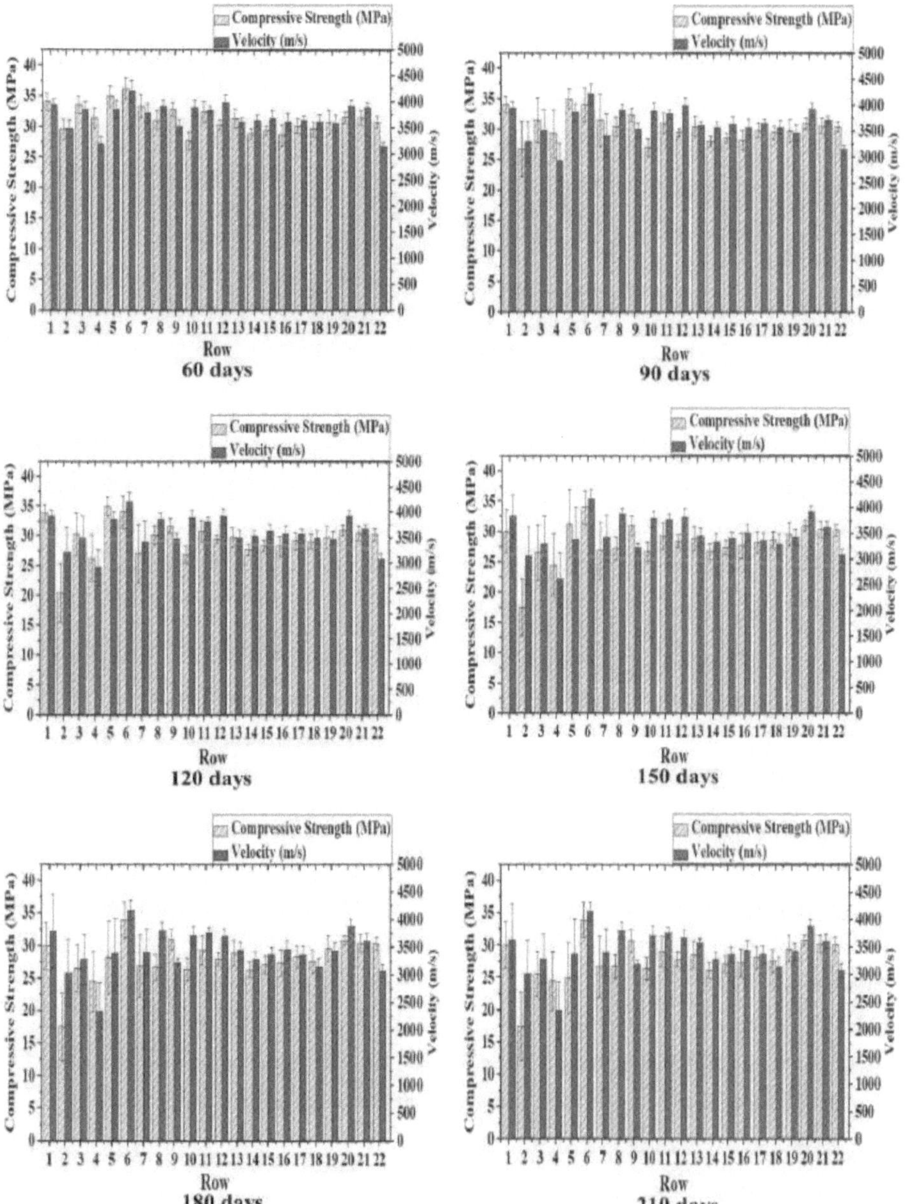


Compressive Strength (MPa)
Velocity (m/s)
Compressive Strength (MPa)
Velocity (m/s)
Compressive Strength (MPa)
Velocity (m/s)
Compressive Strength (MPa)
Velocity (m/s)
Compressive Strength (MPa)
Velocity (m/s)
Compressive Strength (MPa)
Velocity (m/s)
Compressive Strength (MPa)
Velocity (m/s)
Row
60 days
Row
90 days
Row
120 days
Row
150 days
Row
180 days
Row
210 days

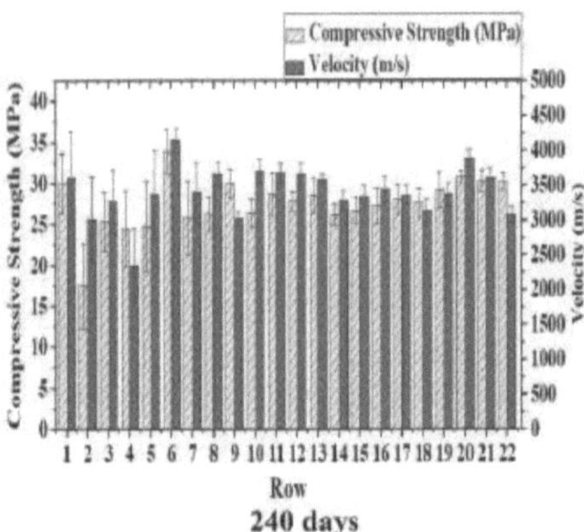


**Figure 44. Average compressive strength and velocity of paver blocks over 240 days of exposure (1-22 rows)**

Figure 44 indicates that the average velocity of GPC (1-8) and PCC (9-22) decreased about 11.6% and 8.3% respectively when the age of exposure increases from 1 day to 240 days. The average compressive strength of GPC (1-8) and PCC (9-22) also decreased about 23.4% and 8.2% over total age of exposure.

According to the American Concrete Institute (ACI) Code 318 [57], the paste content is interrelated to maximum aggregate size. So, GPC paste with small aggregate size desires more amount of AEA. Hence, the authors hypothesize that the scaling of damaged GPC paver blocks was due to an inadequate amount of AEA in the paste.

Permeability is another cause of scaling of concrete subjected to frost action. Although, GPC tends to show a greater permeability resistance than PCC [285], this does not mean low moisture absorption [79]. The general absence of micro-cracks in the Interfacial Transition Zone (ITZ) is the principal reason for low permeability [79]. As can be seen in Figure 45, GPC paver blocks absorbed more water than PCC paver blocks. So, this could be another possible reason that GPC blocks experienced scaling phenomena. This finding is in good-agreement with a performed study by Albitar et al. [286] that the cement-based concrete has lower water absorption and sorptivity than fly-ash based and slag-based GPC due to the capillary mechanism of the pastes.

According to a performed study by Khater et al. [78], the curing regime is critical for the geopolymerization process of aluminosilicate gel which causes high early strength gain.

However, heat-treatment must be applied in a proper way that it makes a supreme condition for the dissolution and precipitation of dissolved silica and alumina species. Overall, in this study, the visual inspection showed that another possible reason for GPC deterioration is shorter curing duration for the paver blocks during the manufacturing stage.

**Figure 45. Rate of moisture absorption of GPC and PCC paver blocks**

Three pits shown with red color (Figure 45) were considered for collecting water samples permeated through paver blocks and estimating the leach-ability of GPC and PCC.

### 8.7.2.2 Compressive strength and velocity of PCC

In this study, the initial average reading of compressive strength and velocity of PCC showed lower values than GPC paver blocks. Although, contrary to common belief, the low compressive strength of concrete does not always lead to low durability. Basically, each 1% of adding AEA in matrix decreases the strength of concrete by approximately 5% [79] and as it is reported [79], non-air entrained concrete with higher strength may show lower frost resistance compared to air entrained concrete due to extra air voids provided to decrease the hydraulic pressure during exposure to low temperature. Moreover, it is reported by Levy et al. [287] that uniform distribution of air bubbles/AEA through cement paste increases the freeze-thaw resistance of concrete. So, it can be concluded that a sufficient amount of AEA was added to the PCC matrix. Hence, this is one of the reasons that scaling phenomena was not observed on the surface of PCC paver blocks.

### 8.7.3   Leach-ability of GPC and PCC paver blocks

A healthy ecosystem and clean environment are essential to the survival of mankind and other organisms. Leaching assessment of materials made by waste by-products is vital to characterize the leaching potential of toxic metals. In this research, water-proof sheet membranes were used in the foundation of a section of paver blocks to prevent penetration of water to the sub-grade level and allow sampling of water that has percolated through the paver blocks and has been in contact with them. Water samples permeated through paver blocks were collected into three pits (Figure 45) to trace the total alkalinity, total hardness, pH, total chlorine, phosphate, copper, ammonia, iron, nitrate and nitrite. Every thirty days, six water samples were collected from each GPC and PCC pits. Figure 46 shows that the pH of GPC and PCC is 7.9 and 6.8 respectively and these values were constant over the total age of exposure. GPC paver blocks show higher pH values than PCC which is attributable to the use of KOH [260]. This finding is in good-agreement with a performed study by Poursaee et al. [288] that Na-based GPC made by slag has a higher pH than PCC due to the presence of alkaline solution in GPC mixture. Furthermore, as it is reported by Izquierdo et al. [203] that heat-treatment of GPC improves the matrix microstructure, decreases the porosity of concrete and decreases the leach-ability of the binder. So, it can be concluded that steam curing improved the microstructure of GPC paver blocks and reduced the amount of heavy metals in water samples.

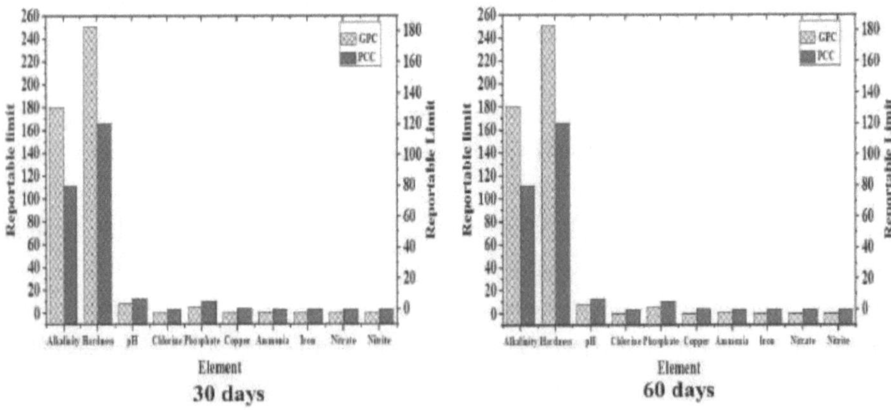

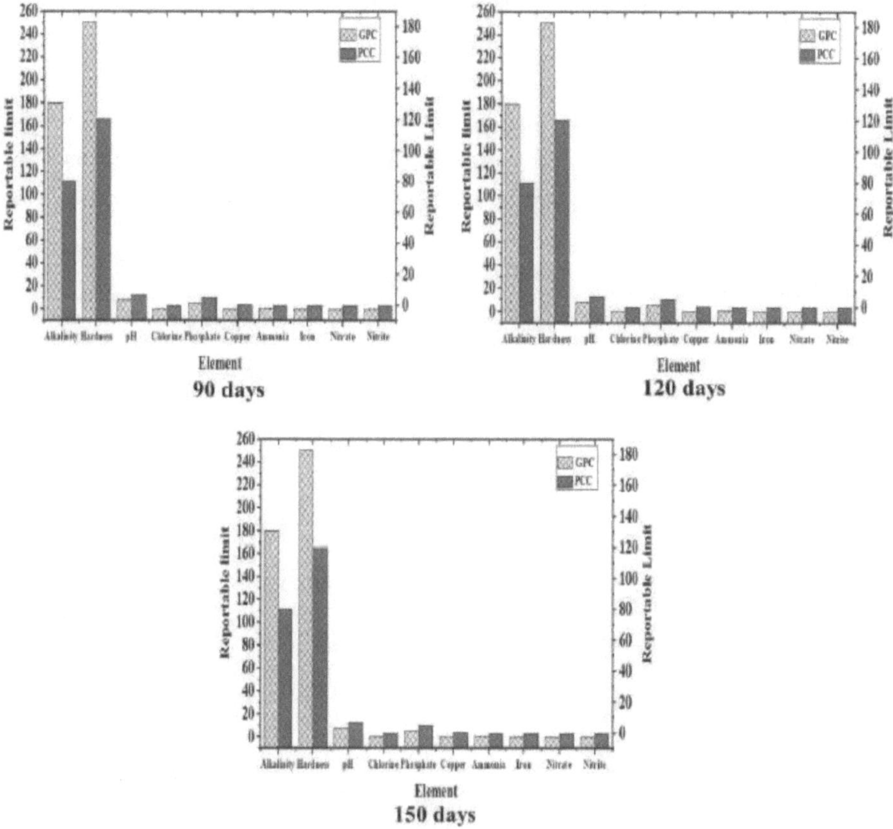



**Figure 46. Leach-ability of chemical elements of GPC and PCC over 150 days of exposure**

Total hardness in water refers to the amount of dissolved calcium (Ca) and magnesium (Mg). Generally, water hardness concentration below 60ppm is considered as soft; 60-120ppm, moderately hard; 120-180ppm, hard, and more than 180ppm, very hard. Kozisek et al. [289] suggested the highest and lowest rate of Ca (40-80mg/l) and Mg (20-30mg/l) in drinking water and defined total hardness as the sum of Ca and Mg concentration of 2-4 mmol/L. The total hardness of GPC and PCC is 120ppm and 180ppm respectively over 150 days of exposure. Iron as one of the main chemical elements of GPC and PCC is 0.15 and 0.2 respectively over the total age of exposure. For both types of concrete, other elements such as chlorine, Nitrate and Nitrite were constant over 150 days and their leach-ability were below the detection limit. The results show that all the metals were immobilized effectively for both types of concretes. All the elements of K-based GPC paver blocks were constant over 150 days as the steam curing method was used to develop the paste structure [203], which is attributed to a full reaction of fly-ash and bottom-ash particles. This is also confirmed by Zhang et al. [290] that slag-based GPC cured at 80 °C for 8 hours can effectively immobilize chemical metals when the amount of chemical metals is in the range of 0.1-0.3%.

### 8.7.4   Dynamic elastic modulus
### 8.7.4.1   Comparison between dynamic elastic modulus of RFT and UPV

In this study, after 180 days of exposure, thirty paver blocks were extracted from the site to make a comparison between RDME derived from UPV and RFT.

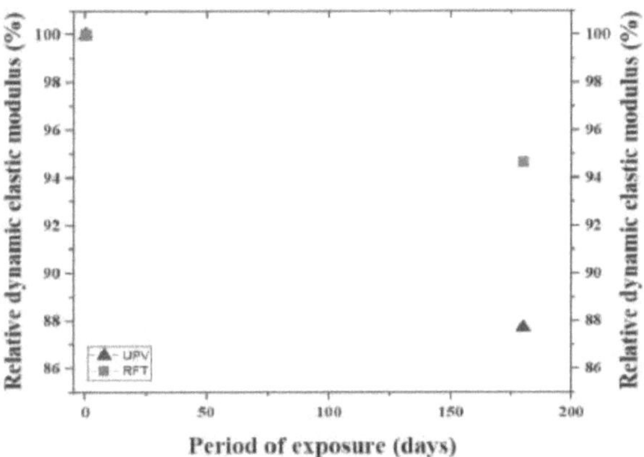


**Figure 47. RDME of UPV vs. RFT**

Average dynamic modulus of elasticity derived from RFT and UPV were calculated in accordance with ASTM C215 [39] and ASTM C597 [275] respectively. In this study, the UPV and the transverse resonant frequency of GPC paver blocks were converted to RDME using Equation (19) and Equation (20) respectively. At zero days of exposure, the average RDME from RFT and UPV was 12.20 GPa and 7.78 GPa respectively. A comparison between dynamic elastic moduli of GPC [36] and reported dynamic elastic moduli for PCC [79] shows that GPC has lower values than PCC. Basically, dynamic elastic moduli of concrete is improved by the modulus of elasticity of the raw constituents, aggregate porosity, properties of paste and characteristics of transition zones such as capillary voids and micro-cracks [79]. The abovementioned finding is confirmed by Fang et al. [121] that due to the various factors including slag replacement level, amount of molarity etc., the dynamic elastic modulus of GPC made by fly-ash and slag is varied from 12 to 58 GPa.

Initially, average RDME of GPC paver blocks before exposure to cold weather was considered to be 100% to have a unique reference and then the average RDME of GPC after 180 days of exposure was calculated and was expressed in percentage to determine material deterioration using the two NDTs.

As can be seen in Figure 47, RDME decreased after 180 days of exposure. It can also be observed that the RDME calculated using the RFT method is higher than that obtained from the UPV method. This difference between the RDME of RFT and UPV can be due to surface conditions among other reasons. About 18 pavers blocks had uneven surface after 180 days of exposure and basically, smoothness of contact the surface affects UPV [291]. While, RFT can measure frequency even on a bumpy surface.

### 8.7.4.2 Dynamic modulus of elasticity of GPC and PCC paver blocks

Table 23 shows the average RDME of GPC and PCC paver blocks placed in area 2. This was calculated using UPV values. As can be seen, the RDME of PCC is higher than that of GPC paver blocks over the entire test period. The RDME decreased progressively in the first two months of exposure for both types of concretes. However, test results reveal that after two months, the RDME of GPC paver blocks decreased much faster than that of PCC.

Table 23. RDME of GPC and PCC over 240 days of exposure

| Exposed (Days) | GPC | PCC |
|---|---|---|
|  | RDME (%) | RDME (%) |
| 1 | 100 | 100 |
| 30 | 98.04 | 99.04 |
| 60 | 94.83 | 97.93 |
| 90 | 94.28 | 96.5 |
| 120 | 91.20 | 96.18 |
| 150 | 88.70 | 95.54 |
| 180 | 88.56 | 95.38 |
| 210 | 88.14 | 95.23 |
| 240 | 85.49 | 95.23 |

Yawei et al. [292] established a damage mechanism model by using RDME and proposed that a power function model is more precise than an exponential function model. The same models were used to create a dynamic elastic modulus attenuation model for both types of paver blocks under real environmental conditions.

Yawei et al. [292] assumed that $E_0$ is the initial dynamic modulus of elasticity of samples and $E_n$ is the dynamic modulus of elasticity after N number of freeze-thaw cycles. In our study, N was considered as exposure period (N=1 for every 30 days) since the temperature variation was not controllable in real environmental conditions. According to the exponential function, RDME attenuation can be created as:

$$Y = E_n/E_0 = ae^{bN} \qquad (21)$$

$$E_n = E_0\, ae^{bN} \qquad (22)$$

The power model also can be used to create a RDME model:

$$Y = E_i/E_0 = aN^b \qquad (23)$$

Where,

Y= RDME of samples, percentage

$E_n$ and $E_i$= Dynamic elasticity modulus of samples after N times freeze-thaw cycles, percentage

$E_0$= Dynamic elasticity modulus of samples before exposing to freeze-thaw cycles, percentage

a and b = Constant variables

The equation (22) and (23) were modeled in Origin software to establish a damage model for both types of concrete used in this study and are plotted in Figure 48(a) and Figure 48(b) respectively. The values from the models are plotted alongside measured field data (shown in Table 23).

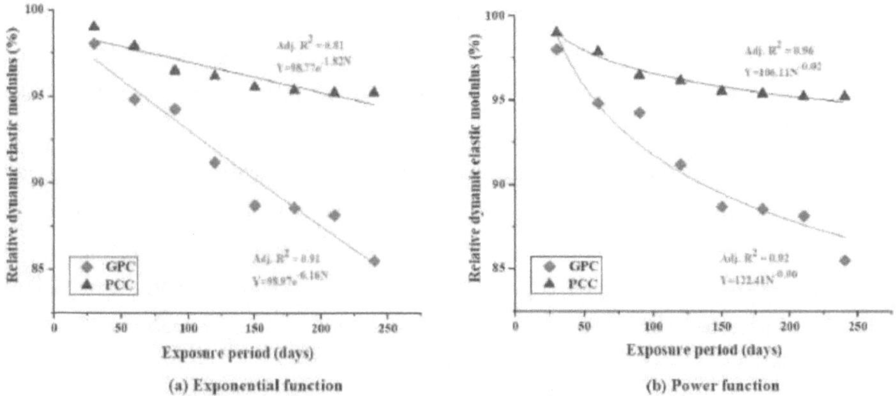



**Figure 48. Comparison between exponential and power function**

According to the power function, the coefficients a and b for GPC were 122.41 and -0.06 respectively. While, the coefficients a and b for PCC were 106.11 and -0.02 respectively. It also can be observed from figures that Adj.$R^2$ of power function of both GPC and PCC is superior compared to the exponential function. Hence, this function can be used to correlate field exposure to damage of GPC and PCC as it has fewer errors. It should, however, be noted that this proposed model is valid for the somewhat mild winter conditions recorded in Victoria, Canada. Hence, caution should be exercised before using this model to paver blocks in more severe weather conditions.

## 8.8 Conclusions

The objectives of this study were to investigate the effect of real environmental exposure on leachability and change in mechanical properties of K-based GPC. This was further compared to traditional concrete (PCC). Numerous conclusions can be drawn from the present study:

- Steam curing and dry curing methods increase the compressive strength of GPC by about 3.5 and 2.3 times respectively when the temperature increases from ambient temperature (-10 °C) to 80 °C. The compressive strength of steam-cured GPC was higher due to the prevention of excessive moisture evaporation, and due to the identical internal curing of GPC samples.

- According to the results of compression test on rectangular and cylindrical GPC samples, rectangular steam-cured GPC samples showed an average compressive strength lower (about 10.28%) than the cylindrical steam-cured GPC samples probably because of the higher rate of stress concentration at corners of rectangular steam-cured GPC samples. The average strength conversion factor between cylindrical and rectangular GPC samples is 0.89.

- The laboratory simulation of seasonal humidity was also carried out to measure the effect of moisture level (dry, SSD, fully wet) on mass and UPV of both GPC and PCC. The results of the experiment indicate that the velocity and mass of GPC and PCC samples increased when humidity level was changed from dry condition to fully wet condition. The mass and velocity ratio of GPC was about 0.98 and 0.84 respectively when the moisture level was increased from dry to fully wet. Whereas, the mass and velocity ratio of PCC samples was 0.97 and 0.79 respectively.

- According to the results of the UPV test, the decrease in velocity of GPC was about 11.6% (area 1) and 12.03% (area 2). While, the velocity of PCC decreased about 8.3% (area 1) and 5.28% (area 2). Moreover, the result of the Schmidt hammer showed that the average compressive strength of GPC decreased about 2.34% (area 1) and 19.51% (area 2). Whereas, the decrease in compressive strength of PCC was about 8.2% (area 1) and 9.65% (area 2).

- The leach-ability of heavy metals, including total alkalinity, total hardness, pH, total chlorine, phosphate, copper, ammonia, iron, nitrate and nitrite of PCC and GPC paver blocks were measured every thirty days (150 days in total) of exposure to real environmental conditions using HACH strips. The results show that the pH of GPC paver blocks is 13.92% higher than pH of PCC paver blocks due to the use of KOH in the GPC paver block mixture. The results also indicated that the steam curing method helped GPC to develop paste structure and caused all the chemical metals of GPC paver blocks to remain constant over 150 days of exposure.

- The RDME was correlated to the field exposure of paver blocks. In this regard, a damage mechanism model was derived for both PCC and GPC paver blocks after 240 days of exposure to real environmental conditions using the power functional model and exponential function model. The comparison between exponential and power function showed that power function is superior to exponential function due to the higher achieved values of Adj.$R^2$ for both GPC (0.96) and PCC (0.92).

## 8.9    Challenges

In developing GPC, the chemical component of precursors including the chemistry phase and size of particles is critical in the reaction process with alkaline solutions. High quality-checking and optimization of coal ashes and activators are always required to make high-quality GPC as the different type of by-product materials are available in various power plants.

Basically, GPC certainly is a sensitive material to temperature and humidity compared to PCC. So, more facilities such as steam curing oven, skills and expertise are required to cure fresh paste in either real environmental conditions or at the laboratory.

Despite the growing market pull for 'sustainable' construction materials, such a technology is not available on an industry-wide level and has restricted applications in the market. Consequently, there are no related standards for GPC reported up to date and basically, standards made for PCC are used for performing tests on GPC.

The curing regime of GPC is another concern in the cold region due to GPC's sensitivity to moisture and temperature. So, in-situ curing in such cold conditions needs more heat for realizing full strength.

## Acknowledgements

The authors appreciatively acknowledge the financial support from Canada-India Research Centre of Excellence (IC-IMPACTS) and authors are also thankful to Dr. Urmil Dave, Professor at Nirma University, for his help in this study; Matt Dalkie, technical services engineer of Lafarge Canada Inc. and Mike McDonald, field chemist of National Silicates an affiliate of PQ Corporation Company for providing materials for this study. Authors also like to thank Priyanka Morla for her help in production of GPC paver blocks.

# Chapter 9   Data in Brief

## 9.1   Flexural strength of GPC

Three GPC specimens were cast to determine the flexural strength of the mixture under three-point loading. Three prism specimens (150×150×553 mm) were cast following ASTM C293 [293] to compare the experimental results with the predictive model for PCC developed by ACI 318 [57] standards and also predictive models developed by Mhaikar et al. [294] and Ahmed et al. [295] to have a better understanding of the flexural behaviour of GPC. The deformed and undeformed shape of beams under three-point loading are shown in Figure 49.

**Table 24. Predictive models**

| Proposed by. | ACI 318 | Mhaikar and Naik | Ahmed et al. |
|---|---|---|---|
| **Citation** | [296] | [294] | [295] |
| 1 | $f_r = 0.62\, f_c^{0.5}$ | $f_r = 0.864\, f_c^{0.5}$ | $f_r = 1.055\, f_c^{0.5}$ |
| 2 | N/A | $f_r = 0.45\, f_c^{2/3}$ | $f_r = \dfrac{0.827}{h^{0.1}}\, f_c^{2/3}$ |

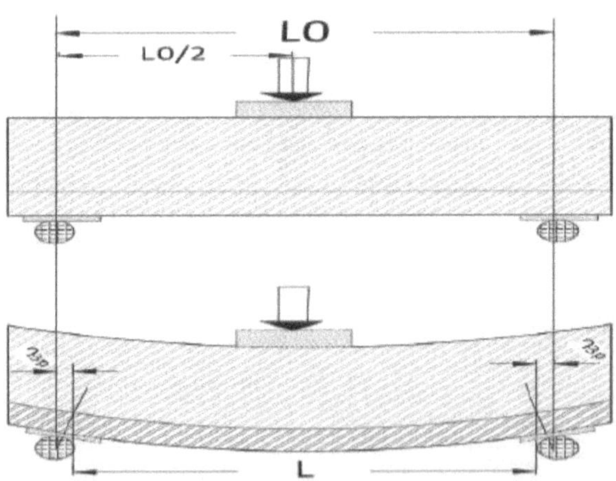


**Figure 49. Three-point loading configuration**

As can be seen in Figure 50, the flexural performance was experimentally measured with a central point bending test. A load frame was equipped with a hydraulic actuator to apply the load at the centre of GPC beam. The GPC prism was supported on metallic rollers. The load was applied with a rate of 1mm/minute.

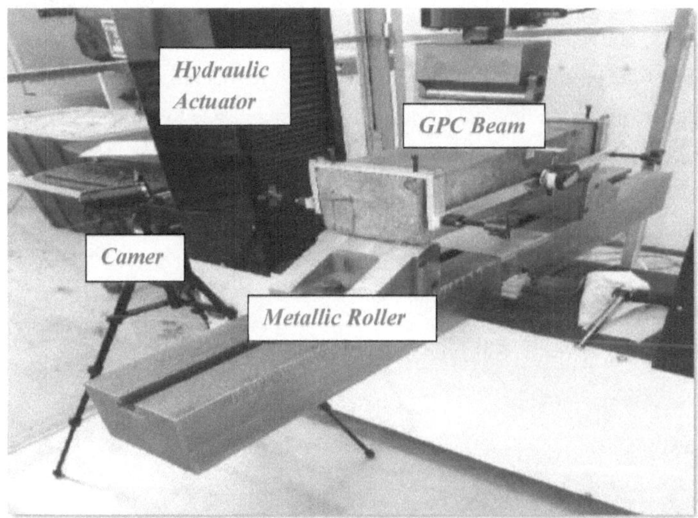


Figure 50. Three-point loading test

NDT was used to find the average compressive strength of each beam. The average compressive strength of ten points on surfaces of GPC samples was measured by using Schmidt hammer following ASTM C805 [274].

The experimental flexural test results and five predictive models proposed by ACI 318 [57] standard [296], Mhaikar et al. [294] and Ahmed et al. [295] were calculated to find an applicable predictive model for GPC. As summarized in Table 25, the flexural strength of GPC specimens increased as compressive strength increases, possibly due to evolution of geopolymerization process and improvement of the microstructure of GPC [3]; thus, GPC becomes denser and consequently, its resistance to crack initiation is expected to be increased [295].

As a comparative study on different flexural strength predictive models, ACI standard model showed relatively close values to experimental test values. The experimental results are about 35% less than ACI standard, because the ACI predictive model is valid for conventional concrete which is less brittle than heat-treated GPC.

Thickness or depth of specimens is one of the important parameters which effects on the ultimate flexural strength of concrete. Ahmed et al. [295] studied the effect of different depth on flexural strength of PCC and used the linear regression analysis to make a predictive model ($A_2$) including depth as one of the parameters. Although, the equation results proposed by Ahmed et al. [295] is about 54% higher than experimental work and it is not applicable for GPC.

**Table 25. Flexural strengtl**

| | Label | GPC 1 | GPC 2 | GPC 3 | Average | Standard Deviation | Coefficient of Variance |
|---|---|---|---|---|---|---|---|
| **f′c (MPa)** | Compressive strength | 29 | 31 | 32 | 30.66 | 1.527 | 0.049 |
| **Experimental Test** | Test | 2.18 | 2.22 | 2.27 | 2.22 | 0.045 | 0.020 |
| **ACI 318** <br> **fr=0.62 f′c0.5** | ACI | 3.33 | 3.45 | 3.50 | 3.42 | 0.087 | 0.025 |
| **Mhaikar et al.** <br> **fr=0.864 f′c0.5** | $M_1$ | 4.65 | 4.81 | 4.88 | 4.78 | 0.117 | 0.024 |
| **Mhaikar et al.** <br> **$f_r=0.45\,f'^{2/3}_c$** | $M_2$ | 4.24 | 4.44 | 4.53 | 4.40 | 0.148 | 0.033 |
| **Ahmed et al.** <br> **$f_r=1.055\,f'^{0.5}_c$** | $A_1$ | 5.68 | 5.87 | 5.96 | 5.83 | 0.142 | 0.024 |
| **Ahmed et al.** <br> **$f_r=\dfrac{0.827}{h^{0.1}}\,f'^{2/3}_c$** | $A_2$ | 4.72 | 4.93 | 5.04 | 4.89 | 0.162 | 0.033 |

**Table 26. The ratio of experimental work with predictive models**

| Sample ID | Test/ACI | Test/ $M_1$ | Test/ $M_2$ | Test/ $A_1$ | Test/ $A_2$ |
|---|---|---|---|---|---|
| **GPC 1** | 0.65 | 0.46 | 0.51 | 0.38 | 0.45 |
| **GPC 2** | 0.62 | 0.45 | 0.48 | 0.36 | 0.44 |
| **GPC 3** | 0.61 | 0.44 | 0.47 | 0.36 | 0.43 |

Figure 51 shows the flexural strength of GPC specimens obtained from six predictive models. The $A_1$ predictive model shows the higher flexural strength values than test values. Hence, it is not applicable for GPC.

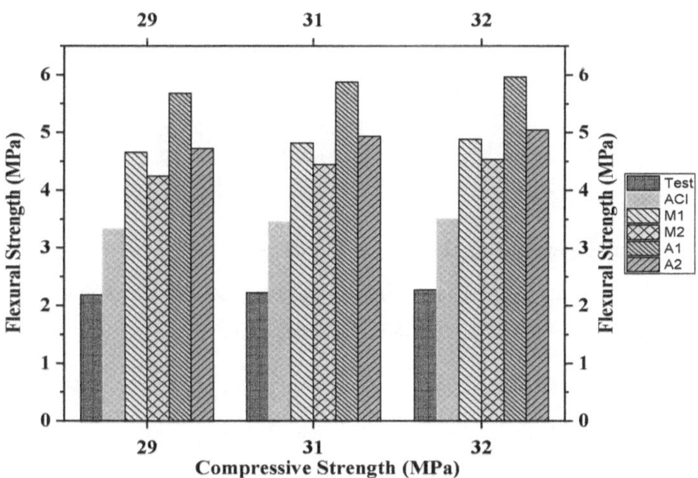


Figure 51. Flexural strength vs. compressive strength of different models

## 9.2 Traffic analysis of paver blocks

Traffic is one of the most significant to quantify factors in paver block design. The deterioration caused to paver blocks by traffic depends on the weight of the vehicles and the number of load repetitions over the traffic analysis period. To quantify this deterioration, the number of Equivalent Single Axle Loads (80 KN ESAL) should be calculated.

IS 15658 [266] recommended different grade of paver blocks for various construction areas and traffic categories. In accordance with IS 15658 [266], the grade of M35 was considered for this study because less than 150 commercial vehicles daily used the parking area. Table 27 shows the total ESALs for this parking site.

**Table 27. ESAL calculation**

| | |
|---|---|
| No. of years to project traffic (assumed) | 2 |
| Average daily traffic | 76 |
| Directional distribution factor | 50 |
| Design lane distribution factor | 100 |
| Growth rate | 2 |
| Percent truck | 2 |
| Truck factor | 0.57 |
| Total ESALs | 319 |

## 9.3 Extra Cost of Steam Curing for $1m^2$:

It is expected that steam curing comes with added costs that include capital cost for equipment (ovens/heaters) and operational costs that include electricity costs.

The lab oven price (Humboldt) =825$

The average price of electricity in Canada per KWh = 0.174$

Average monthly consumption= 1000 KWh

24 hours of steam curing:

1000 KWh/31=32.2 KWh daily

In total:

32.2×0.174=5.61$

# Chapter 10 Conclusions and Future Scope

The research aimed to make K-based GPC using a combination of fly-ash and bottom-ash and investigate its mechanical properties and durability including freeze-thaw resistance, leach-ability, elastic modulus, corrosion etc. using NDTs. In general, the following phases were developed to achieve all proposed objectives:

**Phase I:**

In this phase, a mix proportion for K-based GPC was made after several trial experiments on mortar using various curing regime (ambient, dry and steam). Moreover, the elastic modulus of K-based GPC as one of the most important parameters was measured and compared with the elastic modulus of GPC made by other types of the precursor such as fly-ash.

**Phase II:**

As a novel aspect of this study, micro-level investigation considering the effect of frozen conditions on pore structure of GPC was done. Most of the SEM devices are not able to maintain the temperature of the chamber at the desired temperature. That is why, in this study, 4D-LTSEM equipment as a rare technique was used to capture SEM images from GPC samples at various temperatures including -10 °C, -70 °C, -180 °C. This study was mainly carried out to examine the applicability of ASTM C666 for K-based GPC.

**Phase III:**

One of the main goals of the current study was to investigate the freeze-thaw resistance of this type of GPC. So, accelerating freeze-thaw chamber was used to expose two mix proportions (fly-ash based GPC and bottom-ash based GPC) to 300 freeze-thaw cycles. Moreover, attempts have been made to find a predictive model when GPC is exposed to the freeze-thaw condition. The leach-ability of GPC was measured as another main aim of the current study since it is crucial to find the toxicity of a new material when waste by-products and chemical solution are used. The TCLP test and HACH strips were used to evaluate the potential risk of GPC to release organic and inorganic contaminants into the environment.

**Phase IV:**

First, it should be noted that this phase of research was already defended as a graduate project by another student (Priyanka Morla) at the University of Vitoria. It is well-known that corrosion is initiated when materials that are harmful to steel, such as $CO_2$, start to penetrate concrete and reach the steel reinforcement. That is why, reinforced concrete structures should be tested frequently to detect and prevent corrosion. As these structures get older, the risk of corrosion in reinforcing steel continues to increase. In this phase, two NDTs were used to measure the corrosion resistance of GPC made by fly-ash and bottom-ash since there is limited studies in this area.

**Phase V:**

The characterization of the mechanical properties and durability of GPC specimens and the influence of real environmental conditions entails a specific preparation of the procedures to obtain realistic results as explained in this phase/chapter 8. The NDTs including UPV, HACH strips and Schmidt hammer were applied to estimate the compressive strength, elastic modulus and leaching development of GPC and PCC in real-time considering the practical application (paver block) in situ. Moreover, attempts have been made to create a correlation between freeze-thaw/wet-dry cycles and elastic modulus of paver blocks.

### 10.1 Key findings and accomplishments

For individual phases, the key findings and scientific contributions of the dissertation are briefly presented below:

**Phase I:**

- A new mix proportion was made for producing K-based GPC made by 50% fly-ash and 50% bottom-ash using steam curing method. These results contribute to reducing the emission of $CO_2$ to atmosphere and resource conservation to waste disposal.
- The elastic modulus of this type of concrete was measured and compared with its counterpart PCC using NDTs (RFT/RTG). Moreover, a predictive model was proposed relating elastic modulus to its compressive strength. These results contribute to having a better understanding of the elastic modulus of GPC when a structure/building is designed using this concrete type.

**Phase II**

- The pore structure of GPC under a frozen condition was captured for the very first time in reported literature by 4D-LTSEM. The micrograph of GPC indicated compact and well-bonding microstructure upto -70 °C without any substantial defects. This study provides morphological information about freeze-thaw resistance of K-based GPC and contributes to confirming the applicability of ASTM C666 for GPC.

**Phase III**

- The freeze-thaw durability of two types of GPC (fly-ash based, and bottom-ash based) was evaluated using their RDME. Moreover, the leach-ability of GPC was also measured using TCLP, since chemical activators and waste materials were used in the production of GPC. These results can be very beneficial to find more information about freeze-thaw durability of GPC that can lead to improving this type of concrete's performance under harsh environments. The results of leach-ability contributed to determining that the leached heavy metals from K-based GPC made by bottom-ash and fly-ash are in the range of acceptable standards. Thus, this type of concrete has a less environmental impact and can be used as various applications in the construction sector.

**Phase IV:**

- It is well-known that the corrosion resistance of GPC should be evaluated and compared to traditional concrete to consider it as an alternative material. Unlike PCC, the characteristics of GPC depend upon the raw materials and hence varies for different mixes. Therefore, the corrosion resistance needs to be estimated separately for individual mixes. The proposed methodology and the results add to a better understanding of the corrosion process and provide quantitative information on the performance and effectiveness of GPC which is necessary for evaluating concrete's serviceability and predicting the remaining service life. Moreover, GPC indicated a lower corrosion rate compared to its counterpart PCC because GPC has a moderate to high rate (between 10 μm/year and 20 μm/year). While, PCC showed higher corrosion rate (40 μm/year and 60 μm/year). Furthermore, due to the existence of cracks in PCC samples and penetration of more chloride ions into the paste, PCC showed higher average mass loss (19.15%) compared to GPC (4.14%).

**Phase V:**

- GPC and PCC paver blocks were exposed to an aggressive condition to measure their durability (such as freeze-thaw/wet-dry resistance) and leach-ability using NDTs. A simple empirical model which relates elastic modulus to freeze-thaw/wet/dry cycles was also proposed for both GPC and PCC samples. This phase contributes to knowledge development in the field of GPC by exposing the samples (paver blocks) to real environmental conditions since no study was found to report such criteria.

## 10.2 Recommendations for future work

Based on the results of the present study, recommendations for future work include:

- Basically, the available standards for PCC are currently used for GPC. However, some limitation in some of the standards such as minimum PCC content inhibits their use. Thus, a wide range of work is required to standardize evaluation methods, predictive models and testing procedures related to various parameters of GPC including freeze-thaw, corrosion, and leaching.

- The use of GPC in a wider application similar to that of PCC mostly curing at ambient temperature is a challenging task. The addition of calcium-rich materials such as calcium oxide, calcium hydroxide and granulated blast furnace slag to develop sufficiently high-early-strength of GPC can be useful. Calcium-rich materials accelerate the setting of GPC where additional calcium silicate hydrate (C-S-H) coexists with geopolymer products. Thus, as future work, calcium-based materials can be added to the mixture to produce GPC in ambient temperature.

- Although the 4D-LTSEM method was successful in determining changes in the matrix after sublimating process, more temperatures such as +80 °C (steam-curing temperature of the current study) can also be considered to evaluate the molecular properties and mechanism of GPC.

- As another novel study, the leach-ability and pH of GPC can also be determined using the 4D-LTSEM method at various temperatures.

- The effect of loading on the freeze-thaw durability and their combined effect on serviceability and residual loading capacity is also another important factor that can be studied in future work.

- Developing numerical models for GPC using various software such as COMSOL, which considers various parameters together such as curing, pH, and durability.
- Collecting more data from field paver blocks using different methods/devices such as sensors or strain gauges and correlating them with laboratory data.

# References

[1]     A. Petrillo, R. Cioffi, C. Ferone, F. Colangelo and C. Borrelli, "Eco-sustainable Geopolymer Concrete Blocks Production Process," Agriculture and Agricultural Science Procedia, vol. 8, pp. 408-418, 2016.

[2]     J. Alam and M. Akhtar, "Fly ash utilization in different sectors in Indian scenario," International journal of emerging trends in Engineering and Development, vol. 1, no. 1, 2011.

[3]     J. Davidovits, Geopolymer chemistry and application, 4 ed., France: Institut Geopolymere, 2015.

[4]     T. Silverstrim, H. Rostami, J. Larralde and A. Samadi-Maybodi, "Fly ash cementitious material and method of making a product," US patent, vol. 5, pp. 60-643, 1997.

[5]     J. Van Jaarsveld, J. van Deventer and L. L., "The potential use of geopolymeric materials to immobilize toxic metals: Part I. Theory and Applications," Minerals Engineering, , vol. 7, no. 10, pp. 659-669, 1997.

[6]     F.N.Okoye, J.Durgaprasad and N.B.Singh, "Fly ash/Kaolin based geopolymer green concretes and their mechanical properties," Data in Brief, vol. 5, pp. 739-744, 2015.

[7]     P. R. Vora and U. Dave, "Parametric Studies on Compressive Strength of Geopolymer Concrete," Procedia Engineering , vol. 51, pp. 210-219, 2013.

[8]     F. U. A. Shaikh, "Mechanical and durability properties of fly ash geopolymer concrete containing recycled coarse aggregates," International Journal of Sustainable Built Environment, vol. 5, no. 2, pp. 277-287, 2016.

[9]     ASTM C666, "Standard test method for resistance of concrete to rapid freezing and thawing," ASTM International, America, 2015.

[10]    N. B. Singh, "Fly Ash-Based Geopolymer Binder: A Future Construction Material," Minerals, vol. 8, no. 7, 2018.

[11]    S. Pilehvara, A. M.Szczotoka, J. F. Rodríguez, L. Valentini, M. Lanzón, R. Pamies and A.-L. Kjøniksen, "Effect of freeze-thaw cycles on the mechanical behavior of geopolymer concrete and Portland cement concrete containing micro-encapsulated phase change materials," Construction and Building Materials, vol. 200, pp. 94-103, 2019.

[12]    M. T. Ley, N. Harris, K. J. Folliard and K. C. Hover, "Investigation of air-entraining admixture dosage in fly ash concrete," ACI Materials Journal, vol. 105, no. 5, pp. 494-498, 2008.

[13]    L. Carabba, S. Manzi and M. C. Bignozzi, "Superplasticizer Addition to Carbon Fly Ash Geopolymers Activated at Room Temperature," Materials, p. doi:10.3390/ma9070586, 2016.

[14]    H. Haji-Esmaeili, "Admixtures for Use in Geopolymers," Lakehead University , Ontario, Canada, 2012.

[15]    W. Gwenzi and N. M. Mupatsi, "Evaluation of heavy metal leaching from coal ash-versus conventional concrete monoliths and debris," Waste Management, vol. 49, pp. 114-123, 2016.

[16]    P. Morla and R. Gupta, "Corrosion Resistance of Bottom Ash and Fly Ash-Based Reinforced Geopolymer Concretes Using Half Cell Potential and Linear Polarization Resistance Methods," University of Victoria , Victoria, Canada, 2018.

[17]    Radhakrishna, "Eco-Efficient Masonry Bricks and Blocks," Design, Properties and Durability, pp. 329-358, 2015.

[18]    S. Z. L. Ahmari, "The properties and durability of alkali-activated masonry units," Handbook of Alkali-Activated Cements, Mortars and Concretes, pp. 643-660, 2105.

[19]    K. Pasupathy, M. Berndt, J. Sanjayan, P. Rajeev and D. S. Cheema, "Durability Performance of Precast Fly Ash–Based Geopolymer Concrete under Atmospheric Exposure Conditions," Journal of Materials in Civil Engineering , vol. 30, 2018.

[20]    J. Davidovits, Geopolymer chemistry and application, France : Geopolymere instute, 2015.

[21]    B. Singh, I. G., G. M. and B, "Geopolymer concrete: A review of some recent developments," Construction and Building Materials, vol. 85, pp. 78-90, 2015.

[22]    A. Palomo, M. Grutzeck and M. Blanco, "Alkali-activated fly ashes: a cement for the future," Cement Concrete Research, vol. 29, pp. 1323-1329, 1999.

[23]    R. Rangan, Concrete construction engineering handbook (2nd ed.), New York: CRC Press, 2007.

[24]    H. H and R. RV., Development and properties of low-calcium fly ash based geopolymer concrete, Perth, Australia: Faculty of Engineering, Curtin University of Technology, 2005.

[25]    T. Xie and T. Ozbakkaloglu, "Behavior of low-calcium fly and bottom ash-based geopolymer concrete cured at ambient temperature," Ceramic International , vol. 41, pp. 5945-5958, 2015.

[26]    A.Rajarajeswari and G.Dhinakaran, "Compressive strength of GGBFS based GPC under thermal curing," Construction and Building Materials, vol. 126, pp. 552-559, 2016.

[27]  M. Kaur, J. Singh and M. Kaur, "Synthesis of fly ash based geopolymer mortar considering different concentrations and combinations of alkaline activator solution," Ceramics International, vol. 44, pp. 1534-1537, 2018.

[28]  A. Narayanan and P. Shanmugasundaram, "An Experimental Investigation on Flyash-based Geopolymer Mortar under different curing regime for Thermal Analysis," Energy and Buildings, vol. 138, pp. 539-545, 2017.

[29]  P. Awoyera and A. Adesina, "A critical review on application of alkali activated slag asa sustainable composite binder," Case Studies in Construction Materials, 2019.

[30]  A. Hosan, S. Haque and F. Shaikh, "Compressive behaviour of sodium and potassium activators synthetized flyash geopolymer at elevated temperatures: A comparative study," Journal of Building Engineering, vol. 8, pp. 123-130, 2016.

[31]  A.Palomo, M.W.Grutzeck and M.T.Blanco, "Alkali-activated fly ashes: A cement for the future," Cement and Concrete Research, vol. 29, pp. 1323-1329, 1999.

[32]  D. S. Perera, O. Uchida, E. R. Vance and K. S. Finnie, "Influence of curing schedule on the integrity of geopolymers," Journal of Materials Science, vol. 42, pp. 2099-3106, 2007.

[33]  C.Y.Heah, H.Kamarudin, A. M. AlBakri, M.Binhussain, M.Luqman, I. K. Nizar, C.M.Ruzaidi and Y.M.Liewa, "Effect of Curing Profile on Kaolin-based Geopolymers," Physics Procedia, vol. 22, pp. 305-311, 2011.

[34]  K. EN and A. A., "Effects of curing time and temperature on strength development of inorganic polymeric binder based on natural pozzolan," Journal of Material Science, vol. 44, pp. 3088-3097, 2000.

[35]  G. Kovalchuk, Fernandez-Jimenez and A. Palomo, "Alkali-activated fly ash: Effect of thermal curing conditions on mechanical and microstructural development – Part II," Fuel, vol. 86, pp. 315-322, 2007.

[36]  C. Yang and R. Gupta, "Prediction of the compressive strength from resonant frequency for low-calcium fly ash-based geopolymer concrete," Journal of Materials in Civil Engineering, vol. 30, pp. DOI: 10.1061/(ASCE)MT.1943-5533.0002228, 2018.

[37]  F. Ana, A. Palomo and C. Lopez-Hombrados, "Engineering properties of alkali-activated fly ash," ACI Materials Journa, pp. 106-112, 2006.

[38]  W. I. Khalil, W. A. Abbas and I. F. Nasser, "Dynamic modulus of elasticity of geopolymer lightweight aggregate concrete," 2019.

[39]  ASTM C215, "Standard Test Method for Fundamental Transverse, Longitudinal, and Torsional Resonant Frequencies of Concrete," ASTM, USA, 2016.

[40] R. Zhao, Y. Yuan, Z. Cheng, T. Wen, JianLi, F. Li and Z. J. Ma, "Freeze-thaw resistance of Class F fly ash-based geopolymer concrete," Construction and Building Materials, vol. 222, pp. 474-483, 2019.

[41] F. N. Degirmenci, "FREEZE-THAW AND FIRE RESISTANCE OF GEOPOLYMER MORTAR BASED ON NATURAL AND WASTE POZZOLANS," Ceramics-Silikáty, vol. 62, pp. 41-49, 2018.

[42] G. S. Ryu, Y. B. Lee, K. T. Koh and Y. S. Chung, "The mechanical properties of fly ash-based geopolymer concrete with alkaline activators," Construction and Building Materials, vol. 47, pp. 409-418, 2013.

[43] R. Sathia, G. B. K and S. M., "Durability study of low calcium fly ash geopolymer concrete," In: 3rd ACF international conference, Ho chi minh city, 2008.

[44] D. Reddy, E. JB, S. K and T. A, "Experimental evaluation of the durability of fly ash-based geopolymer concrete in the marine environment," 2011.

[45] M. Olivia and H. Nikraz, "Properties of fly ash geopolymer concrete designed by Taguchi method," Materials & Design , vol. 36, pp. 191-198, 2012.

[46] Y. Xiaolu, "Leaching Behavior of Heavy Metals from Cement Pastes Containing Solid Wastes," 2018.

[47] C. Napia, T. Sinsiri, ChaiJaturapitakkul and P. Chindaprasirt, "Leaching of heavy metals from solidified waste using Portland cement and zeolite as a binder," Waste Management, vol. 32, pp. 1459-1467, 2012.

[48] H. Lu, F. Wei, J. Tang and J. P. Giesy, "Leaching of metals from cement under simulated environmental conditions," Journal of Environmental Management, vol. 169, pp. 319-327, 2016.

[49] E. Arioz, O. Arioz and M. K. O., "Leaching of F-type fly Ash Based Geopolymers," Procedia Engineering, vol. 42, pp. 1114-1120, 2012.

[50] Z. Yunsheng, S. Wei, C. Qianli and C. Lin, "Synthesis and heavy metal immobilization behaviors of slag based geopolymer," Journal of Hazardous Materials, vol. 143, pp. 206-213, 2007.

[51] R. Kolli and M. PothaRaju, "Mechanical Properties of Geopolymer Concrete Composites," Materialstoday Proceedings, vol. 4, pp. 2937-2945, 2017.

[52] H. Jian, Y. Jie, J. Zhang, Y. Yu and G. Zhang, "Synthesis and characterization of red mud and rice husk ash-based geopolymer composites," Cement and Concrete Composites, vol. 37, pp. 108-118, 2013.

[53]    V. Sata, A. Sathonsaowaphak and P. Chindaprasirt, "Resistance of lignite bottom ash geopolymer mortar to sulfate and sulfuric acid attack," Cement and Concrete Composites, vol. 34, no. 5, pp. 700-708, 2012.

[54]    A. Nazari, B. A and R. S., "Properties of geopolymer with seeded fly ash and rice husk barkash," Material Science Engineering, vol. 24, pp. 73-95, 2011.

[55]    P. Nath and P. Kumar, "Flexural strength and elastic modulus of ambient-cured blended low-calcium fly ash geopolymer concrete," Construction and Building Materials , vol. 130, pp. 22-31, 2017.

[56]    AS 3600, "Concrete Structure," Standards Australia , Homebus (NSW Australia), 2009.

[57]    ACI 318, "Building code requirements for structural concrete and commentary," Farmington Hills, MI, USA, 2014.

[58]    S. Kumar, R. Kumar and S. Mehrotra, "Influence of granulated blast furnace slag on the reaction, structure and properties of fly ash based geopolymer," Journal of Material and Science , vol. 45, pp. 607-615, 2010.

[59]    A. Palomo, A. Fernández-Jiménez, G. Kovalchuk, L. M. Ordoñez and M. C. Naranjo, "OPC-fly ash cementitious systems: study of gel binders produced during alkaline hydration," Journal of Materials Science, vol. 42, no. 9, pp. 2958-2966, 2007.

[60]    A. Fernandez-Jimenez, I. García-Lodeiro and A. Palomo, "Durability of alkali-activated fly ash cementitious materials," Journal of Materials Science, vol. 42, no. 9, pp. 3055-3065, 2007.

[61]    E. Kani and A. Allahverdi, "Effects of curing time and temperature on strength development of inorganic polymeric binder based on natural Pozzolan," Journal of Materials Science, vol. 44, pp. 3088-3097, 2009.

[62]    P. Rovnanik, "Effect of curing temperature on the development of hard structure of Metakaolin-based geopolymer," Construction and Building Materials,, vol. 24, pp. 1176-1183, 2010.

[63]    T. L.K and F. Collins, "Carbon dioxide equivalent ($CO_2$) emission: a comparison between geopolymer and OPC cement concrete," Construction and Building Materials, vol. 43, pp. 125-130, 2013.

[64]    E. Gartner, "Industrially interesting approaches to "low $CO_2$" cements," Cement Concrete Res., vol. 34, pp. 1489-1498, 2004.

[65]    S. Wallah, "Creep behavior of fly-ash based geopolymer concrete," Civil Engineering dimension, vol. 12, pp. 73-78, 2010.

[66]    A. Dwivedi and M. Jain, "Fly ash- waste management and overview: A Review," Recent Research in Science and Technology, vol. 6, pp. 30-35, 2014.

[67]    Environmental Protection Agency, "Regulatory Determination on Wastes from the Combustion of Fossil Fuels," Environmental Protection Agency ;, 2000.

[68]    American Coal Ash Association, "Coal Combustion Product Production & Use Survey Report," American Coal Ash Association, Farmington Hills, 2018.

[69]    P. Chindaprasirt, C. Jaturapitakkul, W. Chalee and U. Rattanasak, "Comparative Study on the characteristics of fly-ash and bottom-ash geopolymers," Waste Management, vol. 29, pp. 539-543, 2009.

[70]    D. K. Sinha, A. Kumar and S. Kumar, "Development of Geopolymer Concrete from Fly Ash and Bottom Ash Mixture," Journal of Transactions of the Indian Ceramic Society , vol. 73, pp. 143-148, 2014.

[71]    E. Haq, S. Padmanabhan and AntonioLicciulli, "Synthesis and characteristics of fly ash and bottom ash based geopolymers–A comparative study," Ceramics International, vol. 40, no. 2, pp. 2965-2971, 2014.

[72]    A. F.-J. J. P. G. L. A. P. J. P. Duxson, "Geopolymer technology: the current state of the art," Journal of Material Science , vol. 42, pp. 2917-2933, 2007.

[73]    B. C. McLellan, R. P. Williams, J. Lay, A. V. Riessen and G. D. Corder, "Costs and carbon emissions for geopolymer pastes in comparison to ordinary Portland cement," Journal of Cleaner Production, vol. 19, pp. 1080-1090, 2011.

[74]    G. Habert, J. d. d. Lacaillerie and N. Roussel, "An environmental evaluation of geopolymer based concrete production: reviewing current research trends," Journal of Cleaner Production, vol. 19, pp. 1229-1238, 2011.

[75]    F. Okoye, J. Durgaprasad and N. Singh, "Fly ash/Kaolin based geopolymer green concretes and their mechanical properties," Data in brief, vol. 5, pp. 739-744, 2015.

[76]    X. Y. Zhuang, L. Chen, S. Komarneni, C. H. Zhou, D. S. Tong, H. M. Yang, W. H. Yu and H. Wang, "Fly ash-based geopolymer: clean production, properties and applications," Journal of Cleaner Production, vol. 125, pp. 253-267, 2016.

[77]    P. W. Ken, M. Ramli and C. C. ban, "An overview on the influence of various factors on the properties of geopolymer concrete derived from industrial by-products," Construction and Building Materials, vol. 77, pp. 370-395, 2015.

[78]    H. Khater, "Effect of calcium on geopolymerization of aluminosilicate wastes," J Materials Civil Engineering, vol. 24, pp. 92-101, 2011.

[79]   P. K. Mehta and P. J. Monteiro, Concrete microstructure, properties and materials, 4 ed., MC-Graw Hill, 2014.

[80]   E. A. Sinan and T.Erdoğan, "Alkali activation of a slag at ambient and elevated temperatures," Cement and Concrete Composites, vol. 34, no. 2, pp. 131-139, 2012.

[81]   K.-H. Yang, J.-K. Song and K.-l. Song, "Assessment of $CO_2$ reduction of alkali-activated concrete," Journal of Cleaner Production , vol. 39, pp. 265-272, 2013.

[82]   R. Gupta and H. Rathod, "Current state of K-based geopolymer cements cured at ambient temperature," Emerging materials research , vol. 4, no. 1, pp. 125-129, 2015.

[83]   F. Belforti, P. Azarsa, R. Gupta and U. Dave, "Effect of Freeze-thaw on K-based geopolymer concrete and portland cement concrete," India, 6th Nirma University, 2017.

[84]   C. Yang and R. Gupta, "Prediction of the compressive strength from resonant frequency for low-calcium fly ash-based geopolymer concrete," Journal of Materials in Civil Engineering, pp. DOI:10.1061/(ASCE) MT. 1943-5533.0002228., 2018.

[85]   S. Fathollah, "Effect of curing regime and temperature on the compressive strength ofcement-slagmortars," Construction and Building Materials, vol. 36, pp. 549-556, 2012.

[86]   A. Parghi and M. S. Alam, "Effects of curing regimes on the mechanical properties and durability of polymer-modified mortars – an experimental investigation," Journal of Sustainable Cement-Based Materials, vol. 5, pp. 324-347, 2016.

[87]   C. Heah, H. Kamarudin, A. M. A. Bakri, M. Binhussain, M.Luqman, I. K. Nizar, C. Ruzaidi and Y. Liew, "Effect of Curing Profile on Kaolin-based Geopolymers," 2011 International Conference on Physics Science and Technology (ICPST 2011) , vol. 22, pp. 305-311, 2011.

[88]   G. L. J. P. J. C.K. Yip, "Effect of calcium silicate sources on geopolymerization," Cement Concrete Research , vol. 38, pp. 554-564, 2008.

[89]   A. Noushini and A. Castel, "The effect of heat-curing on transport properties of low-calciumfly ash-based geopolymer concrete," Construction and Building Materials, vol. 112, pp. 464-477, 2016.

[90]   C.Arenas, Y.Luna-Galiano, C.LeivaL.F.Vilches, F.Arroyo, R.Villegas and C.Fernández-Pereira, "Development of a fly ash-based geopolymeric concrete with construction and demolition wastes as aggregates in acoustic barriers," Construction and Building Materials, vol. 134, pp. 433-442, 2017.

[91]    J.Temuujin, A. Riessen and K.J.D.MacKenzie, "Preparation and characterization of fly ash based geopolymer mortars," Construction and Building Materials, vol. 24, pp. 1906-1910, 2010.

[92]    P. Chindaprasirt, T.Chareerat and V.Sirivivatnanonb, "Workability and strength of coarse high calcium fly ash geopolymer," Cement and Concrete Composites, vol. 29, pp. 224-229, 2007.

[93]    N. Lee and H. Lee, "Setting and mechanical properties of alkali-activated fly ash/slag concrete manufactured at room temperature," Construction and Building Materials, vol. 47, pp. 1201-1209, 2013.

[94]    G. Görhan and G. Kürklü, "The influence of the NaOH solution on the properties of the fly ash-based geopolymer mortar cured at different temperatures," Composites Part B: Engineering, vol. 58, pp. 371-377, 2014.

[95]    F. A. Memon, M. F. Nuruddin, S. Demie and N. Shafiq, "Effect of Curing Conditions on Strength of Fly ash-based Self-Compacting Geopolymer Concrete," International Journal of Civil and Environmental Engineering, vol. 5, pp. 342-345, 2011.

[96]    M. F. Nurruddin, S. Haruna, I. G. Sha'aban and B. S. Mohammed, "Methods of curing geopolymer concrete: A review," International Journal of Advanced and Applied Sciences, vol. 5, no. 1, pp. 31-36, 2018.

[97]    ASTM C618, Standard Specification for Coal Fly Ash and Raw or Calcined Natural Pozzolan for Use in Concrete, USA: ASTM, 2015.

[98]    P. Azarsa and R. Gupta, "Novel Approach to Microscopic Characterization of Cryo Formation in Air Voids of Concretes," Micron, vol. 122, pp. 21-27, 2019.

[99]    ASTM C127/C127M, "Standard Test Method for Relative Density (Specific Gravity) and Absorption of Coarse Aggregate," American Society for Testing and Materials, USA, 2015.

[100]   ASTM C33/C33M, "Standard Specification for Concrete Aggregates," American Society for Testing and Materials, USA, 2015.

[101]   ASTM C150/C150M, "Standard Specification for Portland Cement," American Society for Testing and Materials, USA, 2017.

[102]   CSA A23.1, "Standard for concrete materials and methods of concrete construction," Canadian Standard Association , Canada, 2014.

[103]   ASTM C192/C192M, "Standard Practice for Making and Curing Concrete Test Specimens in the Laboratory," American Society for Testing and Materials, USA, 2015.

[104] M. B. Satpute, W. M. R and P. S. V., "Effect of Duration and Temperature of Curing on Compressive Strength of Geopolymer Concrete," International Journal of Engineering and Innovative Technology (IJEIT), vol. 1, no. 5, pp. 152-155, 2012.

[105] ASTM C39/C39M, "Standard Test Method for Compressive Strength of Cylindrical Concrete Specimens," American Society for Testing and Materials, USA, 2014.

[106] D. Hardjito, S. Wallah, D. Sumajouw and B. Rangan, "The stress-strain behaviour of fly-ash based geopolymer concrete," Developments in mechanics of structure and materials , pp. 831-834, 2005.

[107] D. Hardjito and B. Rangan, "Development and properties of low-calcium fly ash-based geopolymer concrete," Perth, 2005.

[108] ACI 363, "State of the art report on high strength concrete," Detriot, MI, USA, 1992.

[109] AS 1012.17, "Determination of the static chord modulus of elasticity and Poisson's ratio of concrete specimens," Australian Standard, Australia, 1997.

[110] B. Tempest, "Engineering characterization of waste derived geopolymer cement concrete for structural applications," Charlotte, 2010.

[111] ASTM C469 / C469M, "Standard Test Method for Static Modulus of Elasticity and Poisson's Ratio of Concrete in Compression," American Society for Testing and Materials, USA, 2002.

[112] w. Prachasaree, S. Limkatanyu, A. Hawa and A. Samakrattakit, "Development of equivalent stress block parameters for fly-ash based geopolymer concrete," Arab Journal Science Engineering , vol. 39, pp. 8549-8558, 2014.

[113] R. Thomas and S. Peethamparan, "Alkali-activated concrete: Engineering properties and stress-strain behaviour," Construction and Building Materials, vol. 93, pp. 49-56, 2015.

[114] A. Wardhono, "The durability of fly-ash geopolymer and alkali-activated slag concrete," Melbourne, Australia, 2015.

[115] T. Ng and S. Foster, "Development of high performance geopolymer concrete. In future in mechanics and structures and materials," London, UK, 2008.

[116] E. Diaz-Loya, E. Allouche and S. Vaidya, "Mechanical properties of fly-ash based geopolymer concrete," ACI materials, vol. 102, pp. 300-306, 2011.

[117] C. M. Gunasekara, "Influence of properties of fly ash from different sources on the mix design and performance of geopolymer concrete," Melbourne, Australia, 2016.

[118] A. C143/C143M, "Standard Test Method for Slump of Hydraulic-Cement Concrete," American Society for Testing and Materials, USA, 2015.

[119] O. K. A. K. J. B. M. Talha Junaid, "A mix design procedure for low calcium alkali activated fly ash-based concretes," Construction and Building Materials , vol. 79, pp. 301-310, 2015.

[120] C.K.Y.Leung, "Concrete as a Building Material," in Encyclopedia of Materials: Science and Technology (Second Edition), ISBN 978-0-08-043152-9, Pergamon, 2001, pp. 1471-1479.

[121] G. Fang, W. K. Ho, W. Tu and M. Zhang, "Workability and mechanical properties of alkali-activated fly ash-slag concrete cured at ambient temperature," Construction and Building Materials, vol. 172, pp. 476-487, 2018.

[122] P. S. Deb, P. Nath and P. K. Sarke, "The effects of ground granulated blast-furnace slag blending with fly ash and activator content on the workability and strength properties of geopolymer concrete cured at ambient temperature," Materials and Design, vol. 62, pp. 32-39, 2014.

[123] S. Pangdaeng, T. Phoo-ngernkham, V. Sata and P. Chindaprasirt, "Influence of curing conditions on properties of high calcium fly ash geopolymer containing Portland cement as additive," Material and Design, vol. 53, pp. 269-275, 2014.

[124] X. Li, Z. Wang and Z. Jiao, "Influence of Curing on the Strength Development of Calcium-Containing Geopolymer Mortar," Materials , vol. 6, pp. 5069-5076, 2013.

[125] Z. Abdollahnejad, A. Dalvand, M. Mastali, T. Luukkonen and M. Illikainen, "Effects of waste ground glass and lime on the crystallinity and strength of geopolymers," Magazine of Concrete Research , p. https://doi.org/10.1680/jmacr.18.00300, 2018.

[126] B. B. Das and N. Neithalath, Sustainable Construction and Building Materials, Springer, 2018.

[127] K. Vijai, R. Kumutha and B. Vishnuram, "Effect of types of curing on strength of geopolymer concrete," International Journal of Physical Science, vol. 5, pp. 1419-1423, 2010.

[128] P. Nath, P. K. Sarker and V. B. Rangan, "Early age properties of low-calcium fly ash geopolymer concrete suitable for ambient curing," Procedia Engineering, vol. 125, pp. 601-607, 2015.

[129] K. Dombrowski, B. A. and M.Weil, "The influence of calcium content on the structure and thermal performance of fly ash based geopolymers," Journal of Materials Science, vol. 42, pp. 3033-3043, 2007.

[130] J. Provis, "Green concrete or red herring? – Future of alkali-activated materials," Advances in Applied Ceramic, vol. 113, pp. 472-477, 2014.

[131] B. V.F.F and J. McKenzie, "Synthesis and thermal behavior of potassium sialate geopolymers," Materials Letters, vol. 57, pp. 1477-1482, 2003.

[132] M. Lizcano, H. S. Kim, S. Basu and M. Radovic, "Mechanical properties of sodium and potassium activated metakaolin-based geopolymers," Journal of Materials Science, vol. 47, pp. 2607-2616, 2011.

[133] A. D. Hounsi, G. Lecomte-Nana, G. Djétéli, P. Blanchart, D. Alowanou, P. Kpelou, K. Napo, G. Tchangbédji and M. Praisler, "How does Na, K alkali metal concentration change the early age structural characteristic of kaolin-based geopolymers," Ceramic International, vol. 40, pp. 8593-8962, 2014.

[134] N. R. Committee, "In-place Strenght Evaluation- A Recommended Practice," NRMCA Publication 133, NRMCA, Silver Spring, MD, 2003.

[135] Indian Standard 456, "Plain and Reinforced Concrete," Indian standard, IIT Kanpur, 2000.

[136] S. P. Pradip Nath, "Use of OPC to improve setting and early strength properties of low calcium fly ash geopolymer concrete cured at room temperature," Cement and Concrete Composites, vol. 55, pp. 205-214, 2015.

[137] F. Hamidi, F. Aslani and A. Valizadeh, "Compressive and tensile strength fracture models for heavyweightgeopolymer concrete," Engineering Fracture Mechanics, vol. 231, p. 107023, 2020.

[138] Z. Abdollahnejad, T. Luukkonen, M. Mastali, P. Kinnunen and M. Illikainen, "Development of One-Part Alkali-Activated Ceramic/Slag Binders Containing Recycled Ceramic Aggregates," Journal of Civil Engineering , vol. 31, pp. DOI: 10.1061/(ASCE)MT.1943-5533.0002608, 2019.

[139] A. El-Newihy, "Application of impact resonance method for evaluation of the dynamic elastic properties of polypropylene fiber reinforced concrete," University of Victoria , Victoria, Canada, 2017.

[140] S. Massoud, J. v. Deventer, P. Mendis and G. Lukey, "Engineering properties of inorganic polymer concretes (IPCs)," Cement and Concrete Research , vol. 37, pp. 251-257, 2007.

[141] A. Fernandez-Jimenez, A. Palomo and C. Lopez-Hombrados, "Engineering properties of alkali-activated fly ash concrete," ACI Materials Journal, vol. 103, pp. 106-112, 2006.

[142] P. Duxson, S. Mallicoat, G. Lukey, W. Kriven and J. v. Deventer, "The effect of alkali and Si/Al ratio on the development of mechanical properties of metakaolin-based

geopolymers," Colloids and Surfaces A: Physicochem. Eng. Aspects , vol. 292, pp. 8-20, 2007.

[143] M. Sofi, J. v. Deventer, P. Mendis and G. Lukey, "Engineering properties of inorganic polymer concretes (IPCs)," Cement and Concrete Research, vol. 37, pp. 251-257, 2007.

[144] M. F. Granata, P. Margiotta and M. Arici, "Simplified Procedure for Evaluating the Effects of Creep and Shrinkage on Prestressed Concrete Girder Bridges and the Application of European and North American Prediction Models," Journal of Bridge Engineering, vol. 18, no. 12, 2013.

[145] C. Argiz, M. A. Sanjuan and E. Menedez, "Coal bottom ash for portland cement production," Advances in Materials Science and Engineering , vol. 2017, no. https://doi.org/10.1155/2017/6068286, p. 7, 2017.

[146] J. Wongpa, K. Kiattlkomol, C. Jaturapitakkul and P. Chindaprasirt, "Compressive strength, modulus of elasticity, and water permeability of inorganic polymer concrete," Materials and design, vol. 31, pp. 4748-4775, 2010.

[147] A. Kar, U. B. Halabe, I. Ray and A. Unnikrishnan, "Nondestructive characterization of alkali activated fly ash and/or slag concrete," European Scientific Journal, vol. 9, pp. 52-74, 2013.

[148] H. J.P, M. P.G. and B. L.T., "Investigation of a synthetic aluminosilicate inorganic polymer," Journal of Material Science , vol. 37, pp. 11-23, 2002.

[149] M. Ankur and R. Siddique, "Strength, permeability and micro-structural characteristics of low-calcium fly ash based geopolymers," Constructions and Building Materials , vol. 141, pp. 325-334, 2017.

[150] 5.-. M. H. Al-Majidi, A. Lampropoulos, A. Cundy and S. Meikle, "Development of geopolymer mortar under ambient temperature for in-situ application," Construction and Building Materials , vol. 120, pp. 198-211, 2016.

[151] P. Azarsa and R. Gupta, "Specimen preparation for nano-scale investigation of cementitious repair material," Micron, vol. 107, pp. 43-45, 2018.

[152] S. Hanjitsuwan, T. Phoo-ngernkham and N. Damrongwiriyanupap, "Comparative study using Portland cement and calcium carbide residuevas a promoter in bottom ash geopolymer mortar," Construction and Building Materials , vol. 133, pp. 128-134, 2017.

[153] T. C. Powers, "The physical structure and engineering propertiesof concrete," Portland Cement Associacition Bulletin, vol. 90, pp. 1-28, 1958.

[154] T. C. Powers and R. A. Helmuth, "Theory of volume changes in hardened Portland cement paste during freezing," Highway Research Board Bulletin, vol. 32, pp. 285-297, 1953.

[155] H. Yu, H. Maa and K. Yan, "An equation for determining freeze-thaw fatigue damage in concrete and a model for predicting the service life," Construction and Building Materials , vol. 137, pp. 104-116, 2017.

[156] M. Kosior-Kazberuka and P. Berkowskib, "Surface scaling resistance of concrete subjected to freeze-thaw cycles and sustained load," Procedia Engineering , vol. 172, pp. 513-520, 2017.

[157] H.S.Wong, A.M.Pappas, R.W.Zimmerman and N.R.Buenfeld, "Effect of entrained air voids on the microstructure and mass transport properties of concrete," Cement and Concrete Research , vol. 41, pp. 1067-1077, 2011.

[158] N. Kabashi, C. Krasniqi, H. Morina and A. Dautaj, "Effect of Air voids in Fresh and Hardening properties of Concrete," Tirana, Albania, 2016.

[159] A. Ziaei-Nia, G.-R. Tadayonfar and H. Eskandari-Naddaf, "Effect of Air Entraining Admixture on Concrete under Temperature Changes in Freeze and Thaw Cycles," Materials Today: Proceedings , vol. 5, pp. 6208-6216, 2018.

[160] H.-S. Shang and T.-H. Yi, "Freeze-Thaw Durability of Air-Entrained Concrete," The Scientific World Journal, vol. 2013, p. 650791, 2013.

[161] P. Sun and H.-C. Wu, "Chemical and freeze-thaw resistance of fly ash-based inorganic mortars," Fuel, vol. 111, pp. 740-745, 2013.

[162] K. Wang, P. J. Monteiro, B. Rubinsky and A. Arav, "Microscopic Study of Ice Propagation in Concrete," Aci Materials Journal, Vols. 93-M42, 1996.

[163] D. J. Corr, P. J. M. Monteiro and J. Bastacky, "Microscopic Characterization of Ice Morphology in Entrained Air Voids," Aci Materials Journal, Vols. 99-M18, 2002.

[164] P. J. M. Monteiro, O. Coussy and D. A. Silva, "Effect of Cryo-Suction and Air Void Transition Layer on Hydraulic Pressure of Freezing Concrete," Aci Materials Journal, Vols. 103-M16, 2006.

[165] J. Yu, X. Li, D. Fleming, Z. Meng, D. Wang and A. Tahmasebia, "Analysis on Characteristics of Fly Ash from Coal Fired Power Stations," Energy Procedia, vol. 17, pp. 3-9, 2012.

[166] A. Mal'chik, S. V. Litovkin and P. V. Rodionov, "Investigations of physicochemical properties of bottom-ash materials for use them as secondary raw materials," Volume 91, 2015.

[167] T. Bakharev, J. Sanjayan and Y.-B. Cheng, "Effect of admixtures on properties of alkali-activated slag concrete," Cement and Concrete Research , vol. 30, pp. 1367-1374, 2000.

[168] ASTM C231/231M, "Standard Test Method for Air Content of Freshly Mixed Concrete by the Pressure Method," American Society for Testing and Materials, West Conshohocken, Pennsylvania, USA, 2015.

[169] J. Isac-García, J. A.Dobado, F. G.Calvo-Flores and H. Martínez-García, "Chapter 10 - Microscale," Experimental Organic Chemistry, 353-370, 2016.

[170] A. A. Patil, H. Chore and P. Dode, "Effect of curing condition on strength of geopolymer concrete," Advances in Concrete Construction, vol. 2, no. 1, pp. 29-37, 2014.

[171] C. Zimring and L. Rathje, Encyclopedia of Consumption and Waste: The Social Science of Garbage, New York, NY, USA: 2nd ed., SAGE, 2012.

[172] B. Miller, Miller, B.G. The future role of coal. In Clean Coal Engineering Technology, 2nd ed., Elsevier, Butterworth-Heinemann, ISBN 978-0-12-811365-3, 2017.

[173] F. R. Environmental Protection Agency, "3. Environmental Protection Agency, Federal Registration, 17 April 2015. Available online: https://www.federalregister.gov/documents/2015/04/17/20150–0257/hazardous-and-solid-waste-management-system-disposal-of-coal-combustion-residuals-from-electric.," 17 April 2015.

[174] M. Marceau, M. Nisbet and M. Geem, "Life Cycle Inventory of Portland Cement Manufacture," Portland Cement Association, Skokie, IL, USA, 2006.

[175] J. Davidovits, "Geopolymer cement: a review," Geopolymer Science and Technics, p. 21, 2013.

[176] T. 6. Silverstrim, H. Rostami, J. Larralde and A. Samadi-Maybodi, "Fly Ash Cementitious Material and Method of Making a Product," U.S. Patent 5,601,643, 1997.

[177] Y. Fu, L. Cai and Y. Wu, "Freeze–thaw cycle test and damage mechanics models of alkali-activated slag concrete," Construction Building Materials, vol. 25, pp. 3144-3148, 2011.

[178] N. Thang, B. Ha, P. Ha, D. Minh, H. Duc and L. Quang, "Leaching behavior and immobilizaion of heavy metals in geopolymer synthesized from red mud and fly ash," Key Engineering Materials, vol. 777, pp. 515-522, 2018.

[179] R. M. Hamidi, Z. Man and K. A. Azizli, "Concentration of NaOH and the Effect on the Properties of Fly Ash Based Geopolymer," Procedia Engineering, vol. 148, pp. 189-193, 2016.

[180] W.-H. Lee, J.-H. Wang, Y.-C. Ding and T.-W. Cheng, "A study on the characteristics and microstructures of GGBS/FA based geopolymer paste and concrete," Construction Building Materials, vol. 211, pp. 807-819, 2019.

[181] H. Y., M. Haoxia and Y. Kun, "An equation for determining freeze-thaw fatigue damage in concrete and a model for predicting the service life," Construction Building Materials, vol. 137, pp. 104-116, 2017.

[182] Z. Mengxuan, G. Zhang, K. Htet, M. Kwon, M. Kwon, Y. Xu and M. Tao, "Freeze-thaw durability of red mud slurry-class F fly ash-based geopolymer: Effect of curing conditions," Construction Building Materials, vol. 215, pp. 381-390, 2019.

[183] K. Sakkas, D.Panias, P.P.Nomikos and A.I.Sofiano, "Potassium based geopolymer for passive fire protection of concrete tunnels linings," Tunnelling and Underground Space Technology, vol. 43, pp. 148-156, 2014.

[184] A. Hosan, SharanyHaque and F. Shaikh, "Comparative study of sodium and potassium based fly-ash geopolymer at elevated temperature," Australia, 2015.

[185] S. Kumaravel, "Development of various curing effect of nominal strength Geopolymer concrete," Journal of Engineering Science Technology, vol. 7, pp. 116-119, 2014.

[186] P. Nath and P. Sarker, "Effect of GGBFS on setting, workability and early strength properties of fly ash geopolymer concrete cured at ambient condition," Construction and Building Materials, vol. 66, pp. 163-171, 2014.

[187] K. Aas-Jalobsen, "Fatigue of Concrete Beams and Columns; Division of Concrete Structure," Trondheim, Norway, 1970.

[188] American Society for Testing and Materials, "Standard Test Method for Slump of Hydraulic-Cement Concrete; ASTM C143/C143M," American Society for Testing and Materials, West Conshohocken, PA, USA, 2015.

[189] J. Temuujin, R. Williams and A. van Riessen, "Effect of mechanical activation of fly ash on the properties of geopolymer cured at ambient temperature," Journal of Material Process Technology, vol. 209, pp. 5276-5280, 2009.

[190] C. Reddy, S. Mohanty and R. Shaik, "Physical, chemical and geotechnical characterization of flyash, bottom ash and municipal solid waste from Telangana State in India," International Journal of Geotechnical Engineering, vol. 9, pp. 1-23, 2018.

[191] P. S., T. Phoo-Ngernkham, V. Sata and P. Chindaprasirt, "Influence of curing conditions on properties of high calcium fly ash geopolymer containing Portland cement as additive," Material Design, vol. 53, pp. 269-275, 2014.

[192] V. Yewale, M. Shirsath and S. Hake, "Evaluation of efficient type of curing for geopolymer concrete," International Journal of New Technology Science Engineering, vol. 3, pp. 10-14, 2016.

[193] D. Hardjito and S. Fung, "Fly ash-based geopolymer mortar incorporating bottom ash," Modern Applied Science, vol. 4, pp. 44-53, 2010.

[194] S. Thokchom, K. Mandal and S. Ghosh, "Effect of Si/Al ratio on performance of fly ash geopolymers at elevated temperature," Arab Journal Science Engineering, vol. 29, pp. 977-989, 2012.

[195] D. Khale and R. Chaudhary, "Mechanism of geopolymerization and factors influencing its development: A review," Journal of Material Science, vol. 42, pp. 729-746, 2007.

[196] B. V. and J. MacKenzie, "Synthesis and thermal behaviour of potassium sialate geopolymers," Material Letter, vol. 57, pp. 1477-1482, 2003.

[197] E. Kamseu, A. Rizzuti, C. Leonelli and D. Perera, "Enhanced thermal stability in K2O-metakaolin-based geopolymer concretes by Al2O3 and SiO2 fillers addition," Journal of Material Science, vol. 45, pp. 1715-1724, 2010.

[198] P. Nath and P. Sarker, "Use of OPC to improve setting and early strength properties of low calcium fly ash geopolymer concrete cured at room temperature," Cement and Concrete Composites, vol. 55, pp. 205-214, 2015.

[199] B. El-Eswed, R. Yousef, M. Alshaaer, I. Hamadneh, S. Al-Gharabli and F. Khalili, "Stabilization/solidification of heavy metals in kaolin/zeolite based geopolymers," International Journal of Mineral Process, vol. 137, pp. 34-42, 2015.

[200] G. Qian, D. Sun and J. Tay, "Immobilization of mercury and zinc in an alkali-activated slag matrix," Journal of Hazard Material, vol. 101, pp. 65-67, 2003.

[201] S. Ahmari and L. Zhang, "Durability and leaching behavior of mine tailings-based geopolymer bricks," Construction and Building Material, vol. 44, pp. 743-750, 2013.

[202] V. Nikolic, M. Komljenovic, N. Marjanovic, Z. Bascarevic and R. Petrovic, "Lead immobilization by geopolymers based on mechanically activated fly ash," Ceramic International, vol. 40, pp. 8479-8488, 2014.

[203] M. Izquierdo, X. Querol, C. Phillipart, D. Antenucci and M. Towler, "the role of open and closed curing conditions on the leaching properties of fly ash-slag-based geopolymer," Journal of Hazardous Materials, vol. 176, pp. 623-628, 2010.

[204]  S. Holmgren, M. Pever and K. Fischer, "Constructing low-carbon futures? Competing storylines in the Estonian energy sector's translation of EU energy goals," Energy Policy, vol. 135, p. 111063, 2019.

[205]  L. Huang, G. Krigsvoll, F. Johansen, Y. Liu and X. Zhang, "Carbon emission of global construction sector," Renewable. Sustainable Energy Reviews, vol. 81, pp. 1906-1916, 2018.

[206]  G. E. &. C. S. Analysis, "IEA. https://www.iea.org/reports/global-energy-$CO_2$-status-report," 2019.

[207]  C. C. W. Hung, S.-C. Hsu and K.-L. Cheng, "Quantifying city-scale carbon emissions of the construction sector based on multi-regional input-output analysis," Resources, Conservation and Recycling, vol. 149, pp. 75-85, 2019.

[208]  X. Zhang and F. Wang, "Hybrid input-output analysis for life-cycle energy consumption and carbon emissions of China's building sector," Building Environment, vol. 104, pp. 188-197, 2016.

[209]  V. Malhotra, "High-Performance High-Volume Fly Ash Concrete," ACI Concrete International, 2002.

[210]  J. Davidovits, "Geopolymers," Journal of Thermal Analaysis, vol. 37, pp. 1633-1656, 1991.

[211]  P. Duxson, A. Fernández-Jiménez, J. L. Provis, G. C. Lukey, A. Palomo and J. S. J. v. Deventer, "Geopolymer technology: the current state of the art," Journal of Material Science, vol. 42, pp. 2917-2933, 2007.

[212]  J. L. Provis, "Modelling the formation of geopolymers," Mar. 2006, Accessed: Mar. 02, 2020. .

[213]  J. L. Provis, P. Duxson, J. S. J. V. Deventer and G. C. Lukey, "The role of mathematical modelling and gel chemistry in advancing geopolymer technology," Chemical Engineering Research and Design, vol. 83, pp. 853-860, 2005.

[214]  J. S. J. V. Deventer, J. L. Provis, P. Duxson and G. C. Lukey, "Reaction mechanisms in the geopolymeric conversion of inorganic waste to useful products," Journal of Hazardous Materials, vol. 139, pp. 506-513, 2007.

[215]  J. v. Jaarsveld and J. S. J. V. Deventer, "Effect of the alkali metal activator on the properties of fly ash-based geopolymers," Industrial and Engineering Chemistry Research, vol. 38, pp. 3932-3941, 1999.

[216] B. Zhang, K. J. MacKenzie and I. W. Brown, "Crystalline phase formation in metakaolinite geopolymers activated with NaOH and sodium silicate," Journal of Materials Science, vol. 44, pp. 4668-4676, 2009.

[217] A. S. D. Vargas, D. C. D. Molin, A. C. Vilela, F. J. D. Silva, B. Pavao and H. Veit, "The effects of Na2O/SiO2 molar ratio, curing temperature and age on compressive strength, morphology and microstructure of alkali-activated fly ash-based geopolymers," Cement and Concrete Composition, vol. 33, pp. 653-660, 2011.

[218] J. G. S. v. Jaarsveld, J. S. J. v. Deventer and G. C. Lukey, "The effect of composition and temperature on the properties of fly ash- and kaolinite-based geopolymers," Chemical Engineering Journal, vol. 89, pp. 63-73, 2002.

[219] J. R. Correia, J. d. Brito and A. S. Pereira, "Effects on concrete durability of using recycled ceramic aggregates," Materials and Structures, vol. 39, pp. 169-177, 2006.

[220] S. Hong, W.-L. Lai and R. Helmerich, "Experimental monitoring of chloride-induced reinforcement corrosion and chloride contamination in concrete with ground-penetrating radar," Structure and Infrastructure Engineering, vol. 11, pp. 15-26, 2015.

[221] Y. P. Asmara, J. P. Siregar, C. Tezara, W. Nurlisa and J. Jamiluddin, "Long term corrosion experiment of steel rebar in fly ash-based geopolymer concrete in NaCl solution," International Journal Corrosion, vol. 2016, 2016.

[222] C. Tennakoon, A. Shayan, J. G. Sanjayan and A. Xu, "Chloride ingress and steel corrosion in geopolymer concrete based on long term tests," Material Design, vol. 116, pp. 287-299, 2017.

[223] M. S. Reddy, P. Dinakar and B. H. Rao, "Mix design development of fly ash and ground granulated blast furnace slag based geopolymer concrete," Journal of Building Engineering, vol. 20, pp. 712-722, 2018.

[224] D. V. Reddy, J.-B. Edouard and K. Sobhan, "Durability of fly ash–based geopolymer structural concrete in the marine environment," Journal of Materials in Civil Engineering, vol. 25, pp. 781-787, 2013.

[225] M. Olivia, "Durability related properties of low calcium fly ash based geopolymer concrete," PhD Thesis, Curtin University, 2011.

[226] P.Chindaprasirt and W.Chalee, "Effect of sodium hydroxide concentration on chloride penetration and steel corrosion of fly ash-based geopolymer concrete under marine site," Construction and Building Materials, vol. 63, pp. 303-310, 2014.

[227] W.-H. Lee, J.-H. Wang, Y.-C. Ding and T.-W. Cheng, "A study on the characteristics and microstructures of GGBS/FA based geopolymer paste and concrete," Construction and Building Materials, vol. 211, pp. 807-813, 2019.

[228] C. Tennakoon, A. Shayan, J. G. Sanjayan and A. Xu, "Chloride ingress and steel corrosion in geopolymer concrete based on long term tests," Materials & Design, vol. 116, pp. 287-299, 2017.

[229] ASTM C876, "Standard Test Method for Half-cell Potentials of Uncoated Reinforcing Steel in Concrete," American Society for Testing and Materials, 1999.

[230] C. Gunasekara, D. Law, S. Bhuiyan, S. Setunge and L. Ward, "Chloride induced corrosion in different fly ash based geopolymer concretes," Construction and Building Materials, vol. 200, pp. 502-513, 2019.

[231] J. M. Miranda, A. Fernández-Jiménez, J. A. González and A. Palomo, "Corrosion resistance in activated fly ash mortars," Cement and Concrete Research, vol. 35, pp. 1210-1217, 2005.

[232] S. Gadve, A. Mukherjee and S. N. Malhotra, "Corrosion of steel reinforcements embedded in FRP wrapped concrete," Construction and Building Materials, vol. 23, pp. 153-161, 2009.

[233] S. G. Millard, K. R. Gowers and J. S. Gill, "Reinforcement corrosion assessment using linear polarisation techniques," Spec. Publ., vol. 128, pp. 373-395, 1991.

[234] A. G59-97, "Standard test method for conducting potentiodynamic polarization resistance measurements," American Society for Testing and Materials, 2014.

[235] F. U. Shaikh, "Effects of alkali solutions on corrosion durability of geopolymer concrete," Advances in Concrete Construction, vol. 2, p. 109, 2014.

[236] M. Al-Azzawi, T. Yu and M. N. S. Hadi, "Factors Affecting the Bond Strength Between the Fly Ash-based Geopolymer Concrete and Steel Reinforcement," Structures, vol. 14, pp. 262-272, 2018.

[237] S. A. Austin, R. Lyons and M. J. Ing, "Electrochemical Behavior of Steel-Reinforced Concrete During Accelerated Corrosion Testing," CORROSION, vol. 60, pp. 203-212, 2004.

[238] L. Bertolini, "Steel corrosion and service life of reinforced concrete structures," Structucture and Infrastructure Engineering, vol. 4, pp. 123-137, 2008.

[239] H. Alanaz, M. Yang, D. Zhang and Z. (. Gao, "Early strength and durability of metakaolin-based geopolymer concrete," Magazine of Concrete Research, vol. 69, pp. 46-54, 2016.

[240] N. H. A. S. Lim, H. Mohammadhosseini, M. M. Tahir, M. Samadi and A. R. M. Sam, "Microstructure and Strength Properties of Mortar Containing Waste Ceramic Nanoparticles," Arabian Journal for Science and Engineering, vol. 43, no. 10, pp. 5305-5313, 2018.

[241] H. Mohammadhosseini, N. H. A. S. Lim, M. M. T. R. Alyousef, H. Alabduljabbar and M. Samadi, "Enhanced performance of green mortar comprising high volume of ceramic waste in aggressive environments," Construction and Building Materials, vol. 212, pp. 607-617, 2019.

[242] G. Dimitriou, P. Savva and M. Petrou, "Enhancing mechanical and durability properties of recycled aggregate concrete," Construction and Building Materials, vol. 158, pp. 228-235, 2018.

[243] H. Mohammadhosseini, A. Abdul and H. E. Abdul, "Influence of palm oil fuel ash on fresh and mechanical properties of self-compacting concrete," Indian Academy of Sciences, vol. 40, pp. 1989-1999, 2015.

[244] F. Okoye, "Geopolymer binder: A veritable alternative to Portland cement," Materialstoday: Proceedings, vol. 4, no. 4, pp. 5599-5604, 2017.

[245] H. Hardjasaputra, M. Cornelia, Y. Gunawan, I. Surjaputra, H. Lie, Rachmansyah and G. P. Ng, "Study of mechanical properties of fly ash-based geopolymer," IOP Conf. Series: Materials Science and Engineering 615, 012009, 2019.

[246] L. M. Zerzouri, S. Alehyen, M. E. Alouani and M. Taibi, "The effect of aggressive environments on the properties of a low calcium fly ash based geopolymer and the ordinary Portland cement pastes," Materials Today: Proceedings , vol. 13, pp. 1169-1177, 2019.

[247] D. S. d. T. Pereira, F. J. daSilva, A. B. R. Porto, V. S. Candido, A. C. R. daSilva, F. D. C. G. Filho and S. N. Monteiro, "Comparative analysis between properties and microstructures of geopolymeric concrete and portland concrete," Journal of Materials Research and Technology, vol. 7, pp. 606-611, 2018.

[248] F. Nabeel, M. NeazSheik and M. N.S.Hadi, "Investigation of engineering properties of normal and high strength fly ash based geopolymer and alkali-activated slag concrete compared to ordinary Portland cement concrete," Construction and Building Materials, vol. 196, pp. 26-42, 2019.

[249] Y. Abualrous, "Characterization of Indian and Canadian Fly ash for use in concrete," Department of Civil Engineering, University of Toronto, Toronto, Canada, 2017.

[250] J. G.Speight, "Chemicals in the Environment," in Reaction Mechanisms in Environmental Engineering, Butterworth-Heinemann, 2018, pp. 43-79.

[251] J. Thaarrini and V. Ramasamy, "Feasibility Studies on Compressive Strength of Ground Coal Ash Geopolymer Mortar," Periodica Polytechnica (civil Engineering), vol. 59, pp. 373-379, 2015.

[252] F.Puertas, B.González-Fonteboa, I.González-Taboada, M.M.Alonso, M.Torres-Carrasco, G.Rojo and F.Martínez-Abella, "Alkali-activated slag concrete: Fresh and hardened behaviour," Cement and concrete composites, vol. 85, pp. 22-31, 2018.

[253] D. Sabitha, J. K. Dattatreya, N. Sakthivel, M. Bhuvaneshwari and S. A. J. Sathik, "Reactivity, workability and strength of potassium versus sodium-activated high volume fly ash-based geopolymers," CURRENT SCIENCE, vol. 103, pp. 1320-1327, 2012.

[254] T. Bakharev, "Geopolymeric materials prepared using class F fly ash and elevated temperature curing," Cement and concrete, vol. 35, no. 6, pp. 1224-1232, 2005.

[255] R. Siddique, "Utilization of Industrial By-products in Concrete," Procedia Engineering, vol. 95, pp. 335-347, 2014.

[256] Y. P. Khandol and D. U. Dave, "Development of fly-ash and bottom-ash based alkali activated paver blocks," Department of civil engineering, institute of technology, Nirma University, Ahmedabad-382481., 2016.

[257] S.R.Hillier, C.M.Sangha, B.A.Plunkett and P.J.Walden, "Long-term leaching of toxic trace metals from Portland cement concrete," Cement and Concrete Research, vol. 29, pp. 515-521, 1999.

[258] H. Lu, F. Wei, J. Tang and J. P.Giesy, "Leaching of metals from cement under simulated environmental Conditions," Journal of Environmental Management, vol. 169, pp. 319-327, 2016.

[259] A.P.Galvín, F.Agrela, J.Ayuso, M.G.Beltrán and A.Barbudo, "Leaching assessment of concrete made of recycled coarse aggregate: Physical and environmental characterisation of aggregates and hardened concrete," Waste Management, vol. 34, pp. 1693-1704, 2014.

[260] M. Izquierdo, X. Querol, J. Davidovits, D. Antenucci, H. Nugteren and C. Fernández-Pereira, "Coal fly ash-slag-based geopolymers: Microstructure and metal leaching," Journal of Hazardous Materials, vol. 166, pp. 561-566, 2009.

[261] S. Pilehvar, V. D. Cao, A. M. Szczotok, M. Carmona, L. Valentini, M. Lanzón, R. Pamies and A.-L. Kjøniksen, "Physical and mechanical properties of fly ash and slag geopolymer concrete containing different types of micro-encapsulated phase change materials," Construction and Building Materials, vol. 173, pp. 28-39, 2018.

[262] R. Zhao, Y. Yuan, Z. Cheng, T. Wen, J. Li, F. Li and Z. J. Ma, "Freeze-thaw resistance of Class F fly ash-based geopolymer concrete," Construction and Building Materials, vol. 222, pp. 474-483, 2019.

[263] Z. Yang, R. Mocadlo, M. Zhao, R. D. Jr, M. Tao and J. Liang, "Preparation of a geopolymer from red mud slurry and class F fly ash and its behavior at elevated temperatures," Construction and Building Materials, vol. 221, pp. 308-317, 2019.

[264] F.-Y. Chang and M.-Y. Wey, "Comparison of the characteristics of bottom and fly ashes generated from various incineration processes," Journal of Hazardous Materials, vol. 138, pp. 594-603, 2006.

[265] X. Goa, B. Yuan, Q. Yu and H. Brouwers, "Characterization and application of municipal solid waste incineration(MSWI) bottom ash and waste granite powder in alkali activated slag," Journal of Cleaner Production, vol. 164, pp. 410-419, 2017.

[266] IS 15658, "Precast Concrete Blocks for Paving- Specification," Indian Standard, India, 2006.

[267] N. Ranjbar and C. Künzel, "Cenospheres: A review," A review, Fuel journal, vol. 207, pp. 1-12, 2017.

[268] F. Goodarzi and H. Sanei, " Plerosphere and its role in reduction of emitted fine fly ash particles from pulverized, coal-fired power plants," Fuel journal, vol. 88, pp. 382-386, 2009.

[269] R. Hooton, "Bridging the gap between research and standards," Cement and Concrete Research, vol. 38, pp. 247-258, 2007.

[270] M. K. Wari, Bajpai, S. Dewangan and U. K, "Fly Ash Utilization: A Brief Review in Indian Context," International Research Journal of Engineering and Technology (IRJET), vol. 03, no. 04, pp. 949-956, 2016.

[271] N. Zabihi and Ö. Eren, "Compressive Strength Conversion Factors of Concrete as Affected by Specimen Shape and Size," Research Journal of Applied Sciences, Engineering and Technology, vol. 7, pp. 4251-4257, 2014.

[272] E. Ferretti, "Shape-Effect in the Effective Laws of Plain and Rubberized Concrete," CMC-Tech Science Press, vol. 30, pp. 237-284, 2012.

[273] EN 206, ""Concrete — Specification, performance, production and conformity," British Standard, UK, 2014.

[274] ASTM C805/805M, "Standard Test Method for Rebound Number of Hardened Concrete," American Society for Testing and Materials, USA, 2013.

[275] ASTM C597/C597M, "Standard Test Method for Pulse Velocity Through Concrete," American Society for Testing and Materials, USA, 2016.

[276] S. A. Omer, R. Demirboga and W. H.Khushefati, "Relationship between compressive strength and UPV of GGBFS based geopolymer mortars exposed to elevated temperatures," Construction and building materials, vol. 94, pp. 189-195, 2015.

[277] AccuaWeather, "AccuaWeather," 2017. [Online]. Available: https://www.accuweather.com/en/ca/victoria/v8r/augustweather/47163?monyr=8/1/2018 &view=table.

[278] K. a. Z. Q. Li, "Influence of freezing rate on damage of cementitious material," Journal of Zhejiang University Science, vol. 10, no. 1, p. 17–21., 2009.

[279] J.Z.Xu, Y.L.Zhou, Q.Chang and H.Q.Qu, "Study on the factors of affecting the immobilization of heavy metals in fly ash-based geopolymers," Materials Letters, vol. 60, pp. 820-822, 2006.

[280] USEPA, "United States Environmental Protection Agency (US-EPA) standard 1311," USEPA, USA, 2016.

[281] H. Khater, "Effect of calcium on geopolymerization of aluminosilicate wastes," Journal of Materials in Civil Engineering, vol. 24, pp. 902-101, 2011.

[282] W. K. Part, M. Ramli and C. C. Ban, "An overview on the influence of various factors on the properties of geopolymer concrete derived from industrial by-products," Construction and Building Materials , vol. 77, pp. 370-395, 2015.

[283] M. Li, H. Hao, Y. Shi and Y. Hao, "Specimen shape and size effects on the concrete compressive strength under static and dynamic tests," Construction and Building Materials, vol. 161, pp. 84-93, 2018.

[284] L. T. Ngoc and C.-A. Graubner, "Uncertainties of concrete parameters in shear capacity calculation of RC members without shear reinforcement," Beton- und Stahlbetonbau, pp. 1-8, 2018.

[285] K. Ramujee and M. Potharaju, "Permeability and abrasion resistance of geopolymer concrete," Indian Concrete Journal, vol. 88, pp. 34-43, 2014.

[286] M. Albitar, M. M. Ali, P. Visintin and M. Drechsler, "Durability evaluation of geopolymer and conventional concretes," Construction and Building Materials, vol. 136, pp. 374-385, 2017.

[287] S. Levy, "Calculations Relating to Concrete and Masonry," in Construction Calculations Manual, 2012, pp. 211-264.

[288] A. Poursaee, " Corrosion of steel in concrete structures," in Corrosion of Steel in Concrete Structures, 2016, pp. 139-33.

[289] F. Kozisek, "Health risks from drinking deminirelized water," Centre of Environmental Health, National Institute of Public Health, Prague, Czech Republic., 2004.

[290] Y. Zhang, S. Wei, C. Qianli and C. Lin, "Synthesis and heavy metal immobilization behaviors of slag based geopolymer," Journal of Hazardous Materials, vol. 143, pp. 206-213, 2007.

[291] Indian Standard 1331, "Non-destructive Testing of Concrete Methods of Test," Indian Standard, India, 2016.

[292] F. Yawei, L. Cai and Y. Wu, "Freeze–thaw cycle test and damage mechanics models of alkali-activated slag concrete," Construction and Building Materials, vol. 25, pp. 3144-3148, 2011.

[293] ASTM C293, "Standard Test Method for Flexural Strength of Concrete (Using Simple Beam with Centre-Point Loading)," American Society for Testing and Materials, USA, 2016.

[294] S. Mhaikar and D. Naik, "Studies on correlation between flexural tensile strength and compressive strength of concrete," The Indian Concrete Journal, vol. 86, pp. 1-6, 2012.

[295] M. Ahmed, K. M. E. Hadi, M. A. Hasan, J. Mallick and A. Ahmed, "Evaluating the co-relationship between concrete flexural tensile strength and compressive strength," International Journal of Structural Engineering, pp. 115-131, 2014.

[296] ACI 318, "Building Code Requirements for Structural Concrete and Commentary," ACI, USA, 2014.

www.ingramcontent.com/pod-product-compliance
Lightning Source LLC
LaVergne TN
LVHW042110190726
843493LV00006B/1437